Naturalizing Epistemology

Also by Fred D'Agostino

CHOMSKY'S SYSTEM OF IDEAS (1986)

FREE PUBLIC REASON (1996)

FREEDOM AND RATIONALITY (*co-edited with* I. C. Jarvie, 1989)

INCOMMENSURABILITY AND COMMENSURATION (2003)

PUBLIC REASON (*co-edited with* Gerald F. Gaus, 1998)

Naturalizing Epistemology

Thomas Kuhn and the 'Essential Tension'

Fred D'Agostino
The University of Queensland, Australia

palgrave
macmillan

First published 2010 by
PALGRAVE MACMILLAN

Palgrave Macmillan in the UK is an imprint of Macmillan Publishers Limited, registered in England, company number 785998, of Houndmills, Basingstoke, Hampshire RG21 6XS.

Palgrave Macmillan in the US is a division of St Martin's Press LLC,
175 Fifth Avenue, New York, NY 10010.

Palgrave Macmillan is the global academic imprint of the above companies and has companies and representatives throughout the world.

ISBN-13: 978–0–230–24099–5 hardback

This book is printed on paper suitable for recycling and made from fully managed and sustained forest sources. Logging, pulping and manufacturing processes are expected to conform to the environmental regulations of the country of origin.

A catalogue record for this book is available from the British Library.

A catalog record for this book is available from the Library of Congress.

10 9 8 7 6 5 4 3 2 1
19 18 17 16 15 14 13 12 11 10

Printed and bound in Great Britain by
CPI Antony Rowe, Chippenham and Eastbourne

This book is dedicated to Shakina Burdo.

Contents

Acknowledgments viii

1 Getting It Together 1

2 Essential Tensions 11

3 The Facilitative Possibilities 20

4 The Consistent Pattern 43

5 Motivating Diversity 64

6 A Culture of Enquiry 81

7 The Division of Labor in Communities of Enquiry 111

8 Some Communities of Enquiry 143

9 Naturalizing the Essential Tension 161

Notes 185

References 204

Index 214

Acknowledgments

The work reported here is part of a project which has received funding support from the Australian Research Council, to which I am extremely grateful. I am grateful too to Steve Fuller for very productive conversation about these matters early in the project and, as always, to my friend Jerry Gaus for sharing his enthusiasms and mine, especially late in the project. I cannot easily capture the degree to which my thinking has been subtly permeated—entirely to my benefit—by the ideas and mode of attack of my friend and sometime colleague, Eric Livingston. Thanks too to editors and referees of the journals *Synthese*, which published a 'headland' paper in which some of the main lineaments of the current volume were sketched (D'Agostino, 2008), *Politics, Philosophy & Economics*, who commissioned and then published (D'Agostino, 2009) work on 'the division of cognitive labor,' which also finds a place here, and *Episteme*, who commissioned and then published work on Kuhn's risk-spreading argument on which I draw heavily, both in substantive terms and motivationally, in attempting the current project. Thanks to my sometime colleagues at the Rijksuniversiteit Groningen, where I was Vonhoff Professor in 2004–5 and began to formulate my thinking on the topics considered here. Thanks, finally, to The University of Queensland, for providing a congenial environment for reflection and for the collective development of my own practical understanding of the pressures and counter-pressures which are involved in developing new initiatives collectively. In this regard, my work on academic leadership, supported by an Australian Learning and Teaching Council grant, is a constant source of exemplary practices and I am grateful to my colleague on this project, Mia O'Brien for all that she has taught me about communities of practice.

1
Getting It Together

> [P]resent scientific growth is unimaginable without institutional means for coordinating the dynamics of research and directing these activities against increasingly explicit problems.
>
> Marlan Blisset, *Politics in Science*

1.1 Introduction

To use some Sixties argot, this is a story about how we 'get it together'... about how, in other words, when we 'get it' about some topic of enquiry—finally understanding how it goes in that domain, we often 'get it' at all only because we have gotten it *together*, that is with the assistance of, in the company of, or against the resistance or opposition of, others. It is, in other words, a we rather than an I that gets it when it is gotten at all, at least in a variety of intellectually and culturally interesting cases, covering both 'normal' and 'revolutionary' periods in the development of traditions, whether they be scientific or organized on some different but nonetheless disciplinary basis.

This general point might be approached metaphysically, as, for instance, Martin Kusch does in Chapter 14 of his recent book *Knowledge by Agreement.*[1] There, Kusch draws on Wittgenstein's so-called private language argument,[2] to argue, specifically,[3] that '[n]ormativity presupposes community' and '[p]ut differently, socially isolated individuals are unable to generate normative phenomena,' or, accordingly, normatively sanctioned claims to knowledge or competence. Richard Rorty represents, perhaps, another, though in his case ironically, metaphysical approach to collective enquiry. Certainly, in 'reducing' objectivity to solidarity,[4] and, more specifically, in the arguments he uses to effect such a reduction, he shows, I think very tellingly, that the advancement of knowledge is unavoidably, rather than merely incidentally, a collective enterprise.

I am not going to approach this general point in this particular way, however. Whatever the interest of the metaphysics and whatever the merits of Kusch's work and Rorty's, which, personally, I find persuasive, there is something even more interesting (in my view).

What is more interesting, I think, are the social dynamics of collective enquiry, the ways in which the collective pursuit of knowledge or competence is facilitated or inhibited by the social dynamics among the individuals concerned.

My story then is in the genre of 'social epistemology,' in the sense made famous in the title of Steve Fuller's book, now 20 years in print.[5] It is also a story, though, about what is called, in other domains, e.g. in Etienne Wenger's influential phrase,[6] the 'community of practice.' I am going to try to tell a story about communities of practice which can generate and test claims about what the world is like. I am going to presuppose that, whether 'socially isolated individuals' *could* do this, as a matter of metaphysics, is largely irrelevant given that this is not the way in which claims typically, perhaps all but universally, *are* generated and tested in contemporary circumstances. So what I am going to do is survey a variety of disciplines including psychology, management science, economics, sociology, anthropology, politics, and philosophy in order to find out *how* communities of enquiry function and how they might function more effectively.

An increasing interest in how we, collectively, generate and test knowledge claims reflects, I think, a 'collective turn' in contemporary thinking across a range of disciplines, a turn which follows the linguistic turn documented by Richard Rorty,[7] the interpretive turn identified by Hiley *et al.*[8] and the practical turn[9] which I sketched, briefly, in my previous book,[10] and which, for me, is best summed up in Pierre Bourdieu's *obiter dictum*:[11] '[t]he social world doesn't work in terms of consciousness; it works in terms of practices, mechanisms, and so forth.' Other important recent contributors to this turn, now threatening, ironically, to become a 'cascade,' are Cass Sunstein, Scott Page, Howard Reingold, and James Surowieckie.[12] I can claim some distinctiveness, despite the range of topics and ideas considered by these authors, at least to the extent that my work, reported herein, makes links to different parts of the scholarly literature and, in particular, ties the whole enterprise to the influential work of Thomas Kuhn, who is largely (and, given his ubiquity at an earlier period, oddly) neglected by these other writers.

I'm going to look, within the collective domain, specifically at how-questions, because, though I will not argue or even document the point here, I believe in their primacy over the sorts of what-questions that have dominated discussions in philosophy. And, while not explicitly arguing for this point, I will try, in the pages which follow, to *exhibit* this primacy, especially by exposing to wider view the fascination of how-questions when they are tackled, as they are by the scholars whose work I am drawing on, in a resolutely empirical manner.

In what follows, I am going, specifically, to consider how groups work on problems and what the potentials and pitfalls are of collective work on problems. I am mindful in this regard that, for example, Francis Bacon famously suggested[13] that groups are, to put it informally, at least partly the problem rather than wholly the solution when it comes to knowing the world. I do not see matters in the same way. While I follow Bacon in acknowledging, indeed insisting on, the particular ways in which groups *can* be a problem—how we can find our own individual intellectual powers undermined by our group allegiances, for example—I also want to highlight the devices, and they are various, that human beings have developed over the centuries to facilitate deliverance of the advantages that groups can bring to the construction and assessment of knowledge. I want, then, to provide a social technology for the collective production of knowledge, where 'technology' has, if you will, specifically Foucauldian overtones.[14]

1.2 Retrospective

In two previous works, *Free Public Reason*[15] and *Incommensurability and Commensuration*, I have argued for an account of 'the right' (whether it be in belief or in normative judgment) that has three main features.

First of all, my account of 'the right' proceeds on a *pluralist* basis. Not only do I recognize that the standards we use to assess claims about 'the right' are diverse and sometimes incommensurable with one another, I insist that this diversity is ineliminable except by means which are themselves obnoxious and that such diversity is in fact a resource as well as (sometimes) a difficulty for decision makers. Accordingly, our account of 'the right' must be diversity-respecting.[16]

Secondly, my account of 'the right' proceeds on a *historicist* basis. In this case, what I recognize is, if you will, the failure of any essentialist account of 'the right' to withstand sustained scrutiny. The concept of 'the right,' in its various manifestations, has a history and its history has a future. There is, in other words, no prospect of finding some empirical or normative basis for making temporally (and perhaps even culturally) universal pronouncements about 'the right.' Furthermore, we should understand that claims about 'the right' are human constructs, in particular settings, and not discoveries of antecedently and independently existing facts about the world, whether of a natural or of some other kind.

Finally, my account of 'the right' proceeds on a *collectivist* basis, in the sense that it recognizes that, inside the social and cultural and attitudinal spaces that they have provided themselves with, individuals are engaged with each other in the historical (and always historically unstable) analysis of 'the right.' Just as there is nothing to be discovered (as essentialism promises) but only to be constructed (as historicism insists), so too is the construction, especially in a pluralist context, always already and henceforth a collective activity.

In *Free Public Reason*, for example, I showed that the standards that we might use to compare the diverse accounts of public reason (of what, in effect, is 'the right' in relation to reasoning publicly about publicly enforceable standards of conduct) are themselves diverse, do not always point in the same direction, may not always be commensurable with one another, but that, historically, groups of women and men might construct for themselves some, always already contestable, account of public reason in terms of which to conduct their collective deliberations. In fact, in that work, I fell back on earlier work in epistemology[17] to present what I called 'the wave model' of public deliberation, according to which it might happen, as it sometimes does during periods of scientific revolution for example, that individuals working in a community setting will collectively bootstrap their way to a properly consensus view about some matters which had hitherto divided them. What one person thinks and commits herself to she may make attractive enough, through her efforts, to enlist the commitment of another who was not, before the innovator's works of improvement, himself sufficiently attracted to that point of view to actually endorse it. Then, with each other to rely on, the two, innovator and disciple, may together make improvements that neither could have made alone, thus attracting favorable attention from some of those whose acceptance threshold was even higher than that of the first disciple. And so on, in a wave-like fashion, with each additional recruit providing the sorts of insights and labor that can, when the wind is in the right direction, eventually produce something which can be *widely* admired and endorsed. Writing later, and with the benefit of literature in social psychology which his own work confirmed me in thinking relevant, the legal theorist Cass Sunstein called[18] this kind of movement of ideas through a group a 'social cascade,' or a 'large-scale social movement . . . in which many people end up thinking something or doing something because of the beliefs or actions of a few "early movers," who greatly influence those who follow.' There are echoes too in Bruce Ackerman's account[19] of 'higher lawmaking' and, especially, of the mobilized deliberation which can result in sweeping constitutional change. Certainly, Thomas Kuhn had something much like it in mind when, much earlier than either Sunstein, Ackerman, or I, he wrote:[20]

> [I]f a paradigm is ever to triumph it must gain some first supporters, men who will develop it to the point where hardheaded arguments can be produced and multiplied. And even these arguments, when they come, are not individually decisive. Because scientists are reasonable men, one or another argument will ultimately persuade many of them. But there is no single argument that can or should persuade them all. Rather than a single group conversion, what occurs is a shift in the distribution of professional allegiances.

In *Incommensurability and Commensuration*, on the other hand, I began to become more completely engaged with the psychological, institutional, and cultural conditions under which collective wisdom could actually be tapped. I noted there, for example, that individuals have to be recruited to and inducted into a community of enquiry before the diversity of their antecedent views and values becomes or even has the potential to become a resource for them, whether individually or collectively. And I considered, drawing especially on the work of the psychologist Philip Tetlock,[21] what particular sorts of beliefs they would have to have about the problem, about their responsibility for its solution, and about the intellectual and other resources of their colleagues in order for it make sense for them to engage in specifically collective deliberations about the issue. Here too my approach was historicist, indeed explicitly so, to the dissatisfaction, of course, of those philosophers (and others) who think of philosophy as a *discovery discipline*. For I had nothing to say about what collective deliberators ought to discover if they deliberate correctly. Indeed, I had nothing concrete and specific to say about how they ought to go about deliberating collectively. I had no normatively binding discoveries about collective deliberation to report, just as, with the earlier book, I had nothing normative to say about what public reason means or, more concretely, what public reasoning might deliver as conclusions about how we ought to live together fairly.

All this attracted a certain amount of bemusement and even consternation, bordering, in less polite circles, on derision. If philosophy can't tell us what reason really is or how it is to be applied in public deliberation or in the evaluation of options, then what good is it? Why bother to consider public reason or evaluation if you can't tell people how they should reason or evaluate alternatives? If philosophy is not a discovery discipline, in my terminology, then what is it and why should anyone want to pursue or engage with it?

I'll return to these questions in Section 9.6. In the meantime...

1.3 Prospective, I

What follows is an argument in eight parts.

In Chapter 2, I return to an important and neglected text by Thomas Kuhn, who, in my view, deserves to be considered the godfather of social epistemology and whose so-called risk-spreading argument constitutes the often unacknowledged basis for much that followed. I point out that one of Kuhn's key topics, and one of his persisting preoccupations, is with the balance, among the individuals constituting a community of practice, between tradition and innovation. Crucially, I remind us that Kuhn addressed this topic in a specifically collectivistic way. He did not think, not really anyway, that any individual enquirer had to achieve some optimal balance between these elements, but, rather, that the community to which individual enquirers might belong had to be so structured that a balance, perhaps shifting from period to period in its history, was maintained at an 'essential tension' between the more traditional or conservative holding fast to what has served well in the past and the more innovative or radical rejection of precisely what has served well in the past (but now promises no longer to do so in perpetuity).

In Chapter 3, I identify some ways in which diversity, within a community of practice, can create opportunities for individuals to enhance their capacities and success in acquiring knowledge or skills. My argument in this chapter is largely functional and relies on well-known observations and 'common sense,' though it does draw, in particular, on some empirical findings from social psychology. I show that a collective approach to knowledge and/or praxis is effective insofar as the work of some individuals

- replicates the work of others, and hence increases the confidence of both in their results;
- complements the work of others, insofar as each works on a part of a larger problems and the solutions found by each can be combined with those found by others to provide insight into the larger problem they are all concerned with;
- extends the work of others by providing alternative interpretations of the problem and of the solution space that ought to be investigated and the techniques of investigation that ought to be used; and
- challenges the work of others and, by doing so, forces the parties to the discussion to consider matters more deeply.

As a preliminary to a discussion of these points, I consider some of the reasons, including especially complexity, contestability, and selectivity, why the capacities of individuals might need to be supplemented by the social organization of labor if we are to generate and evaluate interesting understandings of our world.

In Chapter 4, I look at some empirical (and analytic) findings which suggest that any potential benefits of the 'collectivization of enquiry' will actually be achieved against the 'friction,' if you will, that afflicts or anyway often characterizes group activities. I identify, in that chapter, a number of different frictional elements, among which a generic social-psychological mechanism stands out as of general (and already widely recognized) significance. This is the mechanism of so-called social comparison, on account of which, to put the matter crudely, individuals, anxious about appearing out of step with their colleagues, self-censor, among other things, enunciations of precisely the distinctive knowledge which they have and which, added to the distinctive knowledge of others, might produce what psychologists call an 'assembly bonus'—a bonus we reap from putting the knowledge of one individual together with that of others. Although this generic mechanism is important, partly because it grounds others which play more specific roles in dampening the potential benefits of collectivization, other mechanisms are also important, and are certainly interesting. Among these I might mention now the idea, from economics and social choice theory, of path dependence. Crudely, collective enquiry (like individual enquiry, should it exist in any significant degree) often does not take full advantage of the potential for diversification of efforts because, or to the extent that, individual members of a community of enquiry may find it easier to 'follow the leader' rather than set out on paths that she does not tread. The path that they collectively tread, then, is largely dependent on what the 'first mover' decides, or so, anyway, there is reason to fear. In this chapter, I also consider such interesting issues as the so-called stickiness of knowledge and the ways in which affect, or how enquirers *feel*, can influence their ability to achieve potentially available assembly bonuses.

If the results reported in Chapter 4 provide some ballast against a too-ready enthusiasm for the potential rewards of collectivization, the significance of these results needs to be put into an essential tension, if you will, with other, potentially countervailing results. Hence, in Chapters 5–7, I consider a range of mechanisms which, suitably deployed, might facilitate the delivery of precisely the assembly bonuses which our initial analysis, in Chapter 3, held out hopes for. As Deborah Tollefsen suggests,[22] I will 'look more closely at the context of deliberation in order to uncover the structures that inhibit and promote dissent.' Specifically…

In Chapter 5, I consider a range of motivational factors which, when provided for institutionally, might provide a sufficient counterweight to social comparison and other similar inhibitory mechanisms so that the advantages of collectivization might be more reliably and frequently realized. These factors include, for example, what I call 'multiple accountabilities.' An enquirer who might be inhibited, through social comparison pressures, from sharing her distinctive knowledge with others might be disinhibited if, specifically, she were going to be held to account for not so sharing it by members of another important 'reference group.' Another interesting disinhibitor is what I call the 'premium' which is associated, in certain communities of enquiry, with the novelty of an individual's contribution to collective deliberations and which might provide him with, again, a sufficient counterweight to social comparison pressures, so that he actually does share his distinctive knowledge and/or skills. Finally, I in effect explicate a mechanism already familiar from Kuhn's observations about normal science, namely, that the failure to advance enquiry will be attributed to the individuals whose enquiries were futile and not, at least not normally, to the difficulties inherent in their situation. In psychology, this is known as 'the fundamental attribution error,' but, in communities of enquiry, it can be an important counterweight, as I've put it, to other, inhibitory mechanisms, which might otherwise prevent realization of potential 'assembly bonuses.' (One way in which this study complements the familiar contributions of Cass Sunstein,[23] is, precisely, in identifying the structural and motivational counter-balances to the 'cascades' on which his own work is largely focused.)

In Chapter 6, I consider some 'cultural' aspects of a community of enquiry which might overcome social comparison and other inhibitory pressures working against the realization of potential assembly bonuses. My key point revolves around a pluralistic account of values and a collectivized account of their role in guiding the actions of individuals and of providing the 'cement' that binds them together in a community. While this account reflects my own pluralism (see Section 1.2), it also, I argue, shows Kuhn's approach to normal and revolutionary science in a new, axiologically sophisticated light. Congruent with my belief that Kuhn's fundamental contribution to social epistemology is embodied in his 'risk-spreading argument' (see Section 2.1), I also consider the ways in which the 'demography of risk,' if I may call it that, contributes to the effective collectivization of enquiry. In particular, I consider why it is that we can, typically, rely on individuals within a community of enquiry having, helpfully, a variety of different attitudes toward risk. I show, in effect, that we do not have to posit, but can in fact demonstrate the reality of the 'essential tension' between risk-avoidance and risk-taking. I also consider, in this chapter, familiar institutional features such as agendas, benchmarking, and the importance of attitudes toward risk, in relation to realization of available assembly bonuses.

These Chapters 5 and 6, represent, in effect, the core of my answer to the implied question raised by Scott Page when he said,[24] '[t]he benefits may be there, but we may not know how to unlock them.' I think that we *do* know how to unlock the assembly bonuses potentially available in communities of practice exhibiting internal diversity.

In Chapter 7, I revisit the division of labor and develop a much fuller account of its prospects and limitations. In particular, I bring together the pluralism which I had discussed in Section 3.1 with those ideas about division of labor which I had crudely sketched in Sections 3.2–4. I lean heavily on some recent work in theoretical biology and its adaptations in management science. I look outside philosophy for guidance about the division of labor because, frankly, well-known philosophical discussions of this topic[25] do not really engage with it, but, rather, with another, admittedly related but by no means equivalent topic—namely that of risk and the usefulness of a diversity in attitudes about risk.

In Chapter 8, I revisit some topics raised in earlier chapters, including path dependence, but, more importantly, consider some feedback loops that may provide some understanding of how, despite the complexity of typical domains of enquiry (see Sections 3.1.6 and 7.1.2) and despite the infirmities under which both individuals and collectives suffer (see Sections 2.3 and 3.1) as they try to explore and understand such domains, individuals may contribute precisely to the effective and thorough exploration of the domain, at least up to the usual limits and subject to the usual provisos about entrenchment (Section 4.8) and so on.

Finally, in Chapter 9, I consider some residual issues and some implications for other practices and domains of enquiry. I try to show that the model for a balanced community of enquiry 'scales' up or down to accommodate communities of different sizes and modalities (e.g. face-to-face versus 'virtual' communities). More importantly, I try to develop and deploy some principles for evaluating communities of enquiry, their efficiency and the provisions they make for accountability. I conclude with provisos about the nature of my enquiry. In particular, I defend the approach taken here as providing a *geography* of the issues, rather than a pathway to their solution in some particular way. This is, of course, consistent with, indeed demanded by, my pluralism, my historicism, and, indeed, my understanding of the path-dependence of all of our enquiries. I have had to start somewhere and I will wind up somewhere, but the journey itself is merely illustrative of some aspects of the landscape as a whole. Indeed, there are aspects which are unvisited or even indeed unnoticed by me, so that, even as a map, this volume is always already unfinished.

In all these chapters, topics are considered which I have not mentioned in this brief prospective summary. I have tried, however, to give a sense of both the parts and of their fit with one another.

Next, however, we go back to the future of social epistemology by considering its proximal origins in the work of Thomas Kuhn, one of the most cited and, in my view, most poorly understood, intellectuals of the last century.

2
Essential Tensions

2.1 Kuhn revisited

A common charge, at about the time that Thomas Kuhn's great work *The Structure of Scientific Revolutions* attracted widespread philosophical notice, was that Kuhn was a conservative in relation to science. John Watkins, for example, said[1] that, on Kuhn's account, 'the scientific community [is] an essentially closed society.' According to this interpretation, Kuhn, especially in relation to the idea of normal science, advocated a generally conformist attitude on the part of practicing scientists, interrupted, to make matters worse in the eyes of Kuhn's (largely Popperian) critics, by periodic episodes of alleged collective irrationality—the revolutions of his title. According to Imre Lakatos,[2] for example, 'in Kuhn's view scientific revolution is irrational, a matter for mob psychology.'

This, at that time quite common, interpretation of Kuhn's ideas entirely overlooked his early and continuing concern with the *balance* between conservative *and* innovative dispositions in relation to scientific enquiry; a concern expressed, in his argot, by the phase 'the essential tension,' which named, precisely, the tension between, on the one hand, a conservative commitment to persist in the use of a particular framework of enquiry even in the face of difficulties and, on the other hand, an innovative disposition to look around for new approaches, especially to problems which persistently resist solution along familiar lines. As Kuhn put it:[3]

> Almost none of the research undertaken by even the greatest scientists is designed to be revolutionary, and very little of it has any

> such effect. On the contrary, normal research, even the best of it, is a highly convergent activity based firmly upon a settled consensus acquired from scientific education and reinforced by subsequent life in the profession. Typically, to be sure, this convergent or consensus-bound research ultimately results in revolution. Then, traditional techniques and beliefs are abandoned and replaced by new ones.... [O]nly investigations firmly rooted in the contemporary scientific tradition are likely to break that tradition and give rise to a new one. That is why I speak of an 'essential tension' implicit in scientific research.

Perhaps the most revolutionary of Kuhn's insights, however, was that the solution to the problem of maintaining such an essential tension lies in a *distribution* across *distinct* individuals of *different* balances between conservative and innovative dispositions. Initially, Kuhn made the individual scientist the locus of such an 'essential tension' between conservative and innovative imperatives. As he put it,[4] '[v]ery often the successful scientist must simultaneously display the characteristics of the traditionalist and of the iconoclast.' Even as he said this, though, he already understood, what he later became even clearer about—namely that, as he put it in a footnote,[5] '[s]trictly speaking, it is the professional group rather than the individual scientists that must display both these characteristics simultaneously.... Within the group some individuals may be more traditionalistic, others more iconoclastic, and their contributions may differ accordingly.'

It is not necessarily, in other words, that any given individual enquirer needs to have these tendencies in some optimal balance (or that she needs to oscillate optimally between them) in order for progress in enquiry to be facilitated. It is, rather, that different balances between these tendencies need to be distributed across the community of enquirers (and perhaps shifting balances across time), so that some, more conservative individuals can ensure that the founding concepts or approaches of the community are, as it were, stretched to their limits and not too quickly abandoned when it becomes difficult so to stretch them. *And* that some, more innovative individuals are always (or at least at strategic junctures) harrying the conservatives with new approaches, ideas, and so on, so that, when the founding concepts really can't be stretched any further, there are alternatives available to the community which can be quickly deployed and which can gradually, in a wave-like process,[6] spread out across the entire group, should they prove more promising in relation to the anomalies now encountered.

I said that this was the most revolutionary of Kuhn's insights because, of course, it is, if you will, a founding utterance for a genuinely collective or social approach to epistemology. It is, to be blunt, a feature of the community of enquiry and not of individual members of that community which facilitates the progress of enquiry, according to Kuhn's account. And this makes Kuhn a specifically *social* epistemologist in a very straightforward, and central, sense of that phrase.[7]

Indeed, a related idea of Kuhn's gives more concrete expression to the initial, collectivizing, gesture. This is the idea of risk spreading, introduced already in (the 'Postscript—1969' to) *The Structure of Scientific Revolutions*. As Kuhn said:[8]

> [I]ndividual variability in the application of shared values may serve functions essential to science. The points at which values must be applied are invariably also those at which risks must be taken. Most anomalies are resolved by normal [i.e. conservative] means... If all members of a community responded to each anomaly as a source of crisis..., science would cease. If, on the other hand, no one reacted to anomalies... in high-risk [i.e. innovative] ways, there would be few or no revolutions. In matters like these the resort to shared values rather than to shared rules governing individual choice may be the community's way of distributing risk and assuring the long-term success of its enterprise.

Our first essential tension, then, is that between conservation and innovation, and Kuhn's collectivist approach to this tension has, as I've said, brilliantly pointed the way toward a self-consciously social approach to epistemology. There is, however, a second essential tension and it poses a challenge precisely to Kuhn's analysis of the tension which he so long ago identified and explicated. This second tension is my topic in this work.

2.2 Group dynamics

To give a hint about what is to follow, let me introduce two important ideas, one about the possibly facilitative effects of a collective approach to enquiry, and the other about the inhibitions, in common social-dynamic mechanisms, of precisely such facilitation of enquiry as might seem so readily at hand.

On the one hand, it seems rather obvious that, as Kuhn himself suggests, collectivizing enquiry will produce a bonus over more individualistic approaches. As Kerr *et al.* point out,[9] the collectivization of enquiry can have

> *facilitative possibilities,* including (a) independent parallel processing of information; (b) the elusive 'assembly bonus effect', the combination of different pieces of information that are separately inadequate to produce accurate judgment but which together make a new, emergent solution possible; (c) fellow members catching and correcting one's errors; (d) random error reduction simply though increasing the number of unbiased judgments being integrated; (e) for certain tasks, recognition by group members that an argument or position advocated by another group member is self-evidently correct; or (f) the voicing of alternative positions during group discussion may produce expectancy disconfirmation, which has been shown to undermine judgmental confidence and promote more systematic processing of information.

When individuals have distinctive approaches to their common world, e.g. in striking different balances between conservative and innovative dispositions, then, clearly, they are in a position, collectively, to benefit from this diversity. What one person takes for granted, another challenges, to their mutual benefit; what one person does supplements the distinct activities of others; what one person mistakenly thinks, another person corrects; and so on. In all these cases, there is, in the language of Collins and Guetzkow,[10] a potential 'assembly bonus.'

On the other hand, as social psychologists have been documenting for more than 50 years, collectivization of enquiry can set up social conditions where precisely these bonus effects may not be realized. For example, and summarizing a great deal of research, Hinsz *et al.*[11] point to '[a] *consistent pattern* regarding information processing in groups'—to wit:

> Groups generally *decrease* variability in the way information is processed, compared with individuals. Examples of this pattern include a narrowed focus of attention, redundant memories, accentuation of processing strategies, and shared distribution of information.... Groups are more likely to attend to common (shared) information...and share common information more often than unique information....Shared information may accrue value due to the social validation that arises for information that group members have in common.

In other words, when individuals with different approaches are brought together, as they must be to realize the facilitative possibilities of collectivized enquiry, they become subject, potentially, to social-dynamic pressures which may inhibit their access, collectively, to precisely the diversity which actually exists amongst them and which, if it were tapped, would provide a bonus over individual enquiry. As social comparison theorists have emphasized,[12] individuals have at least two motives when it comes to sharing information or ideas individually arrived at: (a) to be informative and (b) to be accepted in the group in which these information exchanges are occurring.[13] And the latter of these two motives can, all too commonly, interfere with information sharing. So, although the collective *could* be better off given the diversity of individuals' attitudes and approaches than it would be if, as Kuhn's early critics charged, each individual enquirer were simply a conformist to some common standard for attitudes and behavior, it cannot be guaranteed that it *will* be better off given that individuals do not always do what they have to do in order to realize the assembly bonus which their diversity promises. As Stasser and Birchmeier put it,[14] 'the promise of an assembly bonus does not ensure its realization.'

Our second essential tension, then, is between the facilitative possibilities, as Kerr *et al.* call them and the consistent pattern which Hinsz *et al.* have summarized. It is easy to see, for example, how the pressures of social comparison might work well to support dispositions toward conservation, while, on the other hand, the facilitative possibilities have to be realized if innovation or even bare diversity of judgments is to produce real effects. In any event, this is, I think, an empirical analogue of Kuhn's tension, and focusing on it will enable us to continue the work[15] of *naturalizing* social epistemology, or, in other words, of bringing it into contact with evidence, derived from psychology and organization theory, of how, empirically, human beings function within knowledge-producing groups.

I approach this tension, by the way, in the same spirit that Kuhn approached the essential tension between conservation and innovation in science. In particular, while we might think, as Kuhn seems to have thought at first, that the tension between facilitation and inhibition (like that between innovation and conservation) is lodged in the individual, we must acknowledge, as Kuhn did at least parenthetically and then more explicitly, that the tension will be resolved, if at all, only at the social or institutional level—specifically, by identifying the conditions under which inhibitors and facilitators can be balanced for collective advantage, or, as Blisset put it, of identifying those 'institutional means for coordinating the dynamics of research' that enable us to balance these forces effectively.

2.3 Bounded rationality

My project, on this occasion, is situated, I believe, as part of a larger project associated with Herbert Simon, James March, and others.[16] This larger project is encapsulated in the phrase 'bounded rationality,' and what we've learnt from that larger project is that what agents and environments are like, empirically, has something, indeed a great deal, to do with how we, as philosophers (or social scientists), ought to proceed as we chart the terrain that these individuals will traverse in pursuing their aims (see Section 9.6). In the case at hand, the Kuhnian project stands revealed, by worries about social comparison mechanisms, as incomplete, at least in the sense that Kuhn himself did not give an account of how, in the face of these mechanisms, we might actually set up (or at least imagine or understand) a community of enquiry capable of delivering the assembly bonus that we know that science, for example, really has frequently delivered. Kuhn has provided a crucial element: the risk-spreading mechanism of distributed diversity. We now need to supply another crucial element: the institutional and motivational factors, concretely identified, that ensure or at least facilitate the efficacy of this Kuhnian mechanism. Kuhn has sketched what *might* work; our job is to show how what might work can be made to work.

Of course, the consistent pattern that social psychologists have uncovered does not, *per se*, tell against Kuhn's initial analysis. Whether the scientific community, for example, is so much subject to social comparison pressures that assembly bonuses are routinely foregone seems unlikely on the face of it. The history of science, even of normal science, certainly isn't as conspicuously marked as many other enterprises by the reticence of its practitioners to comment on or correct each others' opinions.

What the second, empirical tension *does* show us, though, is that, if scientific (and other) communities of enquiry are *not* hog-tied by social comparison phenomena, then, given how widespread these effects are in *other* important domains (such as legal proceedings, commerce, and politics), there must be other factors in play... factors which it would be useful to identify, precisely to consider whether they might also be deployed in domains where the social comparison pressures more freely influence the interactions of community members and hence limit their prospects for effective enquiry.

Or, to put the matter another way, what the widespread influence of social comparison effects tells us is that, in imagining, designing, or even just seeking to understand a community of enquiry[17] *that actually delivers* an assembly bonus, it will not be enough to identify an appropriate distribution of diversity across the community. Rather, we will also have to identify institutional and motivational (and perhaps recruitment) systems which function to ensure that such social comparison pressures are counteracted and that diversity therefore can play the sort of role which Kuhn envisaged for it.[18]

This sort of 'sophistication,' if I could so style it, is necessary even for the kind of cartography of ideas (see Section 9.6) that I am committed to—of mapping a terrain (rather than, as is more common in philosophy, laying a path to a particular destination). It is even more necessary for those with normative interests, who will need to consider the circumstances in which proposed norms would actually be enacted (or not) in order to be certain of their appropriateness. This is partly a matter of 'uptake,' as I have put it elsewhere[19] in a different context. Unless the parties whose behavior is to be regulated by a proposed set of norms actually (generally) conform their behavior to these norms, the beneficent effects of norm-conforming behavior, which figure significantly in the argument that the norms are actually the right ones, will not in fact occur... and the argument that the norms *are* the right ones will not in fact be cogent. But this is also, at another level, a matter of how we conduct normative enquiries, whether of a specifically philosophical kind or more generally. We need, I submit, to overcome the intellectual snobbery, as Stuart Hampshire put it,[20] 'of abstract thought at the expense of the contingent, of the concrete, of the particular, of the historical accident, of the objects of the presumed lower reaches of the mind' that is often characteristic of the philosophical community, or, more generally, of those who issue pronunciamentos about how people ought to behave. Whether it will be worthwhile to do this will depend, of course, of what additional refinements and what enhanced prospects of uptake we might secure by looking more closely, than is characteristic anyway of many norm-mongers, at the contingent, concrete, particular, and historical circumstances in which knowledge is created and tested. In any event, such close-looking is what I propose to do here.

2.4 Prospective, II

It is my aim, in what follows, to identify institutional, cultural, and motivational systems which might enable enquirers to secure the assembly bonus, and, in so doing, to redeem the promise of what I take to be the fundamental Kuhnian epistemic mechanism (of risk-spreading via distribution of diversity). Like Karin Knorr Cetina,[21] 'I am not interested in the construction of knowledge but in the construction of the machineries of knowledge construction.' And, also like her,[22] 'I will have something to say . . . on the architectures—and the diversity—of the manufacturing systems from which truth effects arise—on the empirical strategies of these systems, on their configuration of objects, technologies, and epistemic subjects . . . .' This is, if you will, the 'cartography' that I discuss in Section 9.6.

Specifically, I will, in what follows, consider three crucial mechanisms for overcoming the inhibiting effects of social comparison (and other similar) mechanisms in communities of enquiry. These are: multiple accountabilities and novelty premiums (Chapter 5); the shallow consensus of communities of enquiry (Chapter 6); and some environmental and institutional factors (Chapter 8) which trigger or at least support a facilitative attitude on the part of members of these communities.

In particular, I want to consider how social comparison pressures might be counterbalanced by what Shadish *et al.*,[23] following Merton, call 'sociological ambivalence.' I also want to consider the relevance of the distinction between intellective and attitudinal diversities.[24] In the course of my exposition, I will revisit my analysis[25] of the *shallow consensus* that characterizes members of a community bound together by their mutual commitment to a paradigm of enquiry.

I will also consider, crucially, the sort of '[h]ighly dynamic, complexly looped'[26] environment of enquiry in which scientists, and other enquirers, typically function, and show how the dynamism and complexity of this environment is itself actually a facilitative condition for successful enquiry. What emerges from this analysis is an argument that the ways in which human enquiry are organized may sometimes trade, in a positive, rather than a negative, way upon the difficulties of conducting enquiry. Precisely the specifically environmental challenges which might hamper some forms of enquiry sometimes in fact support effective enquiry by triggering attitudes which work against common social comparison mechanisms. Conditions which might seem to set limits on enquiry in fact elicit the development of a mode of enquiry which makes even these conditions more amenable to investigation. It is, crudely, the difficulties of doing science that sometimes make it possible for science to overcome the consistent pattern. There is, in the cartographic idiom, sometimes a pathway across the terrain defined by inhibitors and facilitators that leads enquirers to productive engagement with their domains.

Overall, then, I will follow Pierre Bourdieu[27] in addressing the question:

> What are the social conditions which must be fulfilled in order for a social play of forces to be set up in which the true idea is endowed with strength because those who have a share in it have an interest in truth, instead of having, as in other games, the truth which suits their interests?

And, like him,[28] I will consider how, concretely,

> the inherent logic of the field, and in particular, the struggle between the dominant and the new entrants, with the resultant cross-control, ... bring[s] about, under certain conditions, a *systematic diversion of ends* whereby the pursuit of private scientific interests (again in both senses of the word) continuously operates to the advantage of the progress of science.

This is my approach to cashing out the Kuhnian legacy.

3
The Facilitative Possibilities

3.1 More on bounded rationality

I have already mentioned 'bounded rationality.' (See Section 2.3.) This idea is sometimes presented as if it were, in some mildly pejorative sense, merely a 'practical' issue... as if the fact that we cannot, 'in real time,' consider all the alternatives and/or all the evidence bearing on the choice between alternatives, was merely a matter of highly contingent limitations on our processing powers that might in fact be overcome and would in any event be addressed in the case of really important decisions or choices where more and better resources could and would be deployed.

This somewhat belittling account of bounded rationality is not mine, nor is it correct. In fact, we are precluded, in a wide variety of cases, from considering all the options and all the data not merely 'practically,' but, in fact, 'in principle.' Or, in other words, even if we had more time and more resources, we still couldn't consider all the options and all the data. To do so would exceed the powers of *any finite being*.[1] The failure to do so is not merely a contingent fact about us, as the particular kinds of beings that we are. It is, rather, a fact about any being that is not infinite in her powers—and, by the way, it doesn't disappear simply by collectivizing enquiry (see Section 3.1.7). I will discuss this issue under seven headings, as follows:

- Inexhaustibility
- Fundamental uncertainty
- Undecidability
- Reflexivity
- Contestability

- Complexity
- The burdens of judgment

The point of this excursus is, of course, to identify both the necessity for and the limitations on the collectivization of enquiry. Insofar as individual human beings are limited in their cognitive powers, for any or all of the reasons which I will outline, they will see, or at least there will be, some potential point in their joining together to ameliorate these limitations. On the other hand, insofar as some of these limits are not overcome simply by the collectivization of enquiry, then they represent limits too on a more collectivized modality of engagement. They must, in either case, be part of our naturalizing story about enquiry.

3.1.1 Inexhaustibility

Why is there more to bounded rationality than 'practical' limitations on our 'real-time' powers of observation and deliberation? Stuart Hampshire gives forceful and colorful expression to one reason why reason is bounded. He expresses this in terms of 'the inexhaustibility of descriptions,' or the fact, as he puts it,[2] that '[a]ny situation which confronts me, and which is not a situation in a game, has an inexhaustible set of discriminable features over and above... those which are mentionable within the vocabulary that I possess and use.' Indeed, the situation is perhaps even more dire than Hampshire's particular formulation would suggest. For there is, as a simple thought experiment will easily verify, 'an inexhaustible set of discriminable features' in any situation even *within* the limits of the particular vocabulary which we possess and use. Here is an example which I have used in classroom discussions of this point.

An instruction to the class, let it be as large as you like. 'Write down ten features of the current situation which might be relevant to your understanding of and behavior in this situation.' When we assemble the students' answers, one answer which we *never* find is this: 'The distance between the lecturer and the drinks container that has been left on the counter near the lectern.' This is a discriminable feature of the situation, expressible in our shared vocabulary, and it might well be relevant to understanding the situation and its possibilities—for example, if, as contemporary circumstances make vividly relevant, the drinks container contained an explosive device and if the lecturer's prospects of surviving the blast, when the device was ignited, depended, precisely, on the distance between it and him.

I do not think that the demonstration is either fanciful or misleading. We can add to its force, however, by considering another response we never get from students. 'The relative positions, on a multi-dimensional color-chart, of the various patches of color which can be observed both on students' clothing and on the decorative surfaces of furniture and of the room itself.' Even art history students don't give this answer or anything like it, although, at least stripped of technicalities, color terminology is readily available as part of the everyday speech of ordinary folk. Furthermore, as a few moments' reflection will reveal, these sorts of facts might be quite relevant to the role of art director on a motion picture set—they might in fact be precisely the most important or anyway among the most important kinds of facts which he might consider and deliberate about.

These points are articulated, abstractly, by Christopher McMahon, who writes:[3]

> It may be that all the parties translate facts into reasons in the same way, but there are so many facts relevant to the question being considered that it is impossible for anyone to keep them all in mind. In such a case, there can be disagreement because each is focusing on different facts, yet exchanging factual information will not solve the problem because as new information is brought to mind, old information inevitably slips from view.

Even without exceeding the limits of our current vocabularies, then, there is, demonstrably, much more of potential relevance in ordinary everyday situations than we can possibly encompass, in real time, in our thinking. Even, indeed, when we divide the labor of devising situation-relevant descriptions, as in my classroom thought experiment, we never get anything remotely like all, or even indeed any very substantial portion, of the descriptions that actually might be relevant. If this is an infirmity of finite beings and of finite collections of them, it is a very deeply embedded one (which doesn't, by the way, depend on our finitude).

3.1.2 Fundamental uncertainty

David Dequech has drawn attention to a limitation on our observations and deliberations that depends, in effect, on two inescapable facts about the human condition: (a) that human deliberation often involves future states of affairs (as in planning) and (b) that future states of affairs are often 'radically novel,' especially in involving possibilities which are unpredictable at the time of deliberation. As he puts it,[4] using the phrase 'fundamental uncertainty' as a label, 'unimagined and unimaginable new states may occur in the future....' He continues: 'Thus, no optimal solution [for typical problems] exists ex ante that can be objectively defined, even by an imaginary external analyst who does not have the computational limitations of [everyday] decision makers.'

It is easy enough, I think, to see what Dequech has in mind here. If are planning now, at t_0, a course of action that we will take over some future period t_1–t_2, then, unavoidably, our understanding of the environment and opportunities for action during t_1–t_2 will be limited to what, at t_0, we can predict and/or imagine. But there are very good inductive grounds for believing that, on certain matters, our understanding at t_0 of planning-relevant environment and opportunities during t_1–t_2 will often be extremely poor. Most 1970s forward business planning, even in major corporations, would have been notably deficient in anticipating the relevance to business activities in even the 1990s, let alone the 2000s, of portable or mobile information and communication technologies of kinds that were then, in Dequech's simple diction, 'unimaginable'—e.g. the now nearly ubiquitous Blackberry (or iPhone) with its internet access and push e-mail. (Notice, by the way, that, as with inexhaustibility, it is not just individuals who are infirm in this way. Multiplying individuals may ameliorate the problem, but it does not eliminate it.)

What all this means, as Dequech clearly sees, is that there are deliberation-relevant descriptions of the environment and opportunities for action that are not even available at the time of deliberation and, hence, that, as he puts it, 'no optimal solution exists ex ante that can be objectively defined'—the *ex post* optimal solution may well involve descriptions that have yet to be devised at the time planning occurs.

3.1.3 Undecidability

We owe to formal logic the demonstration that the question of theoremhood can be 'undecidable' for certain propositions in certain logical systems.[5] We owe to Stuart Kauffman a model which, using the lambda calculus apparatus, enables us to generalize the idea of undecidability to more general (and less formalized) contexts. Kauffman is concerned with what he calls cultural evolution, which might certainly encompass the development of thinking in reasonably complex areas of enquiry such as science and the humanities. He poses the issue in the form of a (probably false but nonetheless illuminating) dichotomy. As he says:[6]

> Now, if formal undecidability can arise from the real laws of chemistry [when they are represented in the lambda calculus and treated as a formal system], might the same undecidability not arise in technological or even cultural evolution? Either we can capture the unknown laws of technological complementarity and substitutability in some kind of formal grammar or we cannot. If we can, then Gödel's theorem suggests that there will be statements about how such a world evolves that are formally undecidable. And if we cannot, if there are no laws governing the transformations, then surely we cannot predict.

The point is nicely expressed, but will benefit from some commentary.

Consider the collection of statements which are in play in some field of enquiry. Here are some possibilities for describing this collection in a formally rigorous way: (a) The collection is closed under various inferential operations in the sense that taking any combination of statements and applying a suitable operator to that combination, we get a statement which is also a member of the collection of statements; (b) The collection is less well defined in the sense that the set of inferential operations in play among the community of enquiry which generates and assesses this collection is not itself a well-defined set—some inferential operations are widely perhaps even universally accepted, but others are more controversial; and (c) The collection of statements itself is not well defined, in the sense that, while some statements are widely perhaps even universally accepted as belonging to the collection and some are widely perhaps even universally accepted as definitively not belonging to the collection, the membership status of other statements is less widely agreed or less definitively settled.

What Kauffman is pointing to, I think, is the possibility that, if the collection of statements in a given field of enquiry is closed, then, if it is reasonably comprehensive, and if the inferential operations upon that collection are reasonably powerful, there will be reason, in the various undecidability results from formal logic, for thinking that there will be statements of the same general form as statements which are incontrovertibly members of the collection whose membership status with respect to that collection is provably undecidable. There will be a statement S#, for instance, for which we can *demonstrate* that its status is indeterminate with respect to the collection C. This, then, is an in-principle upper (or is it lower?) bound on our rational deliberative powers—we simply cannot say for S# either that it is or that it is not a member of C, for important values of S# and C.

On the other hand, where there is, as in cases (b) and (c) above, some looseness, if you will, in the very formulation of the discursive space, there will, accordingly, be some looseness about what can and what cannot reasonably be said that is in accordance with (or follows from or wouldn't be incompatible with) what has already been said. But, in this case, we are, in effect, no better off than in cases of proven undecidability; for important values of the S# and C variables, we simply don't, antecedently, know what to say about their relations. Of course, in this case, perhaps more obviously than in the case of formal undecidability, the uncertainty may be merely *pro tem*. Perhaps further deliberations and investigations will enable us to settle issues which we cannot now settle. But this too represents an impediment to present-time planning about future-time situations. (See Section 3.1.2.) Again, our powers of deliberation and decision are limited relative to our 'need to know.' And, again, the defect is not a remediable one. In the case of formal undecidability (the first tine on Kauffman's fork), the defect is remediable only by conventional fiat which merely shifts the undecidability elsewhere in the larger system. In the case of ill-definition, if I may so style it, the defect is remediable via ordinary enquiry and debate, but, since decision and action must sometimes be planned in advance of the ill-definition being overcome, and since this is a problem, like the poor, that is always with us as temporally embodied beings, the implications are no less telling in this case, where, as Kauffman puts it, 'surely we cannot predict.'

3.1.4 Reflexivity

Compounding the difficulties posed by inexhaustibility and 'fundamental' uncertainty, we have another factor well known to philosophers of the social sciences for decades,[7] but recently given a 'French twist' if I could put it that way.[8] This factor is typically discussed under the heading 'reflexivity,' and impinges on planning and deliberation, whether individual or collective, as follows.

If are trying now, at t_0, to understand or predict how agents will act over some future period t_1–t_2, we will use and apply a vocabulary at t_0 to characterize these agents and their potential future situation and options that may well contain terms, call them $T_1 \ldots T_m$, not likely to be used by the agents themselves. Insofar as our deliberations become public, however, some of the claims we make may well thereby become available to the agents whose behavior we are considering. But the availability of our claims may well alter how the agents themselves behave, thereby defeating the point of our prediction (or interpretation), whether because the prediction has thereby become self-defeating or, though less obviously, because it has become self-fulfilling.

Some examples will help. Suppose that we point out that agents in a particular situation are likely, because of what they believe about and desire in that situation, to reach a sub-optimal outcome in that situation. Insofar as our analysis becomes a matter of public knowledge, then, if the agents themselves accept it as veridical, they may well so adapt their beliefs and desires (and hence their actions) to defeat the prediction precisely by reaching an outcome that is not or is not as badly sub-optimal. Our analysis has interacted with their deliberations to change their behavior to behavior which falsifies our analysis. This is self-defeating reflexivity. Similar examples can easily be generated of analyses whose dissemination to agents leads those agents to fulfill precisely the terms of the analysis (when, at least with non-zero probability, they might not otherwise have done so).

All this is obvious and familiar. What it shows in the present context is an in-principle limitation on our deliberative powers. Insofar as we are concerned with understanding what people will think, feel, and do in the future, then, insofar as our analyses are parts of the same worlds as their actions and attitudes, there can be no guarantee that our analyses are adequate to their actions and attitudes. And the limitation is not one which is overcome by making more extravagant assumptions about the powers of human beings.

3.1.5 Contestability

With both contestability and complexity, in a sense soon to be explicated, our powers of reasoning are bounded or limited because of the multiplicity of different and not always well-aligned factors which are relevant to our understanding, deliberation, or decision making. Some concepts are contestable, in W.B. Gallie's famous characterization,[9] because of the diversity of different factors which are relevant to their applicability and, crucially, because of indeterminacy in the relations between these factors—there is, to adopt economistic jargon, no settled or consensual or given 'trade off' between the various factors, so that some individuals might be perfectly reasonable, given their own situations, to trade the factors off in one way while other individuals trade them off differently. And if there is (a) no higher standard to adjudicate the differences between these parties, and (b) the need, or even simply the perceived need, to arrive at some binding and consensual approach to the trade-offs, then there will be, inevitably, contestation about the definition and application of these concepts and hence limits to the powers of reason to settle issues which involve them.[10]

Here, I think, we have a particularly vivid example—though not one which all readers will accept as presented—of a limit on our rational deliberations which is grounded in the nature of things, if you will, and not in the mere infirmity of our powers (though there is, in fact, nothing trivial about whatever infirmities may also exist). On the account which I have myself presented in different fora and which is most famously and influentially embodied in the work of Isaiah Berlin,[11] the difficulty in deliberating in the face of contestable concepts is not epistemological—and, *a fortiori*, it is not a matter of our infirmities; it is, rather, ontological—the lack is not in us but in the fabric of the world, or, at least, in the fabric of our values *per se*. John Gray expressed this very clearly:[12]

> [Values] are many, they often come into conflict with one another and are uncombinable in a single human being or a single society, and... in many such conflicts *there is* no overarching standard whereby the competing claims of such ultimate values are rationally arbitratable.

The crucial phrase, to which I have added emphasis, is, of course, the phrase 'there is,' with its ontological import clearly displayed in its logical form. In the absence—not the obscurity, but the literal absence—of an 'overarching standard,' there can be, at best, contestability of claims in cases where there are multiple values or criteria and no settled way of trading them off against one another. And the claim really is, at least in Berlin's case (and mine), an ontological one. As Berlin insists, in the face of such contestability, we should not, as he puts it,[13] 'automatically attribute it to some deficiency on our part which could be eliminated by an increase in skill or knowledge'; it is simply there, as I've put it, in the fabric of our values.

3.1.6 Complexity

If contestability depends on an ineliminable plurality of deliberation-relevant factors (e.g. values in a decision context), then complexity, in the sense that I have in mind, also depends on this sort of plurality. Luigi Marengo and his colleagues provide a very vivid expression, in a rather different idiom from any we have encountered so far (but none the worse for that), of the sorts of difficulties which complexity of certain kinds can pose for the deliberator. As they put it:[14]

> In problems whose solutions involves the exploration of high-dimensional spaces, agents endowed with limited computational capabilities and with a limited knowledge of the interdependencies can explore only a subset, possibly very small, of solutions. Even if we assume that the selection mechanism which selects among alternative solutions works perfectly..., the outcome of selection is bound by the set of solutions produced by the generative mechanism. It may well be the case that optimal or even 'good' solutions will never be generated at all, and thus that they will never be selected by any selection mechanism whatsoever.
>
> The problem here is that strong interdependencies create a large set of local optima in the search space. Marginal contributions rapidly switch from positive to negative values, depending on which value is assumed by other components. As a consequence, the presence of strong interdependencies prevents the possibility of reaching optimal solutions by simply adopting an optimal value for each of the components a problem is made of.

This will bear and benefit from explication.

Imagine, to make matters concrete, that we are trying to decide on a new car to purchase. If we are reasonably assiduous consumers, we may consider a number of different evaluative dimensions $\delta_1 \ldots \delta_n$ over which to rank the various options $O_1 \ldots O_m$. This gives us an $n+1 \times m$ matrix, as follows, with values for the various options relative to the various values as indicated and with δ^* representing overall value.

	δ_1	δ_2	δ_3	...	δ_n	δ^*
O_1	$V_{1/1}$	$V_{1/2}$	$V_{1/3}$		$V_{1/n}$	V_1^*
O_2	$V_{2/1}$	$V_{2/2}$	$V_{2/3}$		$V_{2/n}$	V_2^*
O_3	$V_{3/1}$	$V_{3/2}$	$V_{3/3}$		$V_{3/n}$	V_3^*
...						
O_m	$V_{m/1}$	$V_{m/2}$	$V_{m/3}$		$V_{m/n}$	V_m^*

The point that Marengo *et al.* are making is that δ^* evaluations of the options can vary in quite complex ways relative to variations in underlying dimensions of value $\delta_1 \ldots \delta_n$. So, for example, the overall value of an option might, other things being equal, increase as its δ_1 value increases but only up to a certain threshold $T(\delta_1)$, after which overall value decreases (other things being held constant) as δ_1 value increases.[15] Or perhaps, when δ_1 and δ_2 vary together, there are quite complex patterns of change in the overall value of options. When this is the case, we will get, if we were to map the option-space $m \times n$ dimensionally, a very complex surface where, in particular, overall value might vary, across the surface, in a complexly 'bumpy' manner (rather than across smooth gradients), as illustrated below.[16]

Points which are near one another in an underlying value dimension may be quite far apart when it comes to their δ^* ranking. And this, as Marengo *et al.* point out, may make it quite difficult, to find the best option overall, to search such a 'fitness landscape,'[17] to adopt some biological terminology. Suppose that we were to begin our search for the overall best option from a point at the left-hand end of Figure 3.1 and that we set out toward the right. We would quickly arrive, as overall value continually increased, at the point marked A and, if we proceeded further toward the right, we would quickly discover that overall value now decreased steadily. We might well conclude that point A represents the global maximum of overall value, but, as is clear from our 'bird's-eye view' of the fitness landscape, this maximum is merely a local one. If we were to multiply the already apparent complexity by adding numerous other dimensions, as would certainly be appropriate for many 'real-life' decision problems, we would quickly appreciate the force, reworded slightly, of the claim of Marengo *et al.*,[18] that where

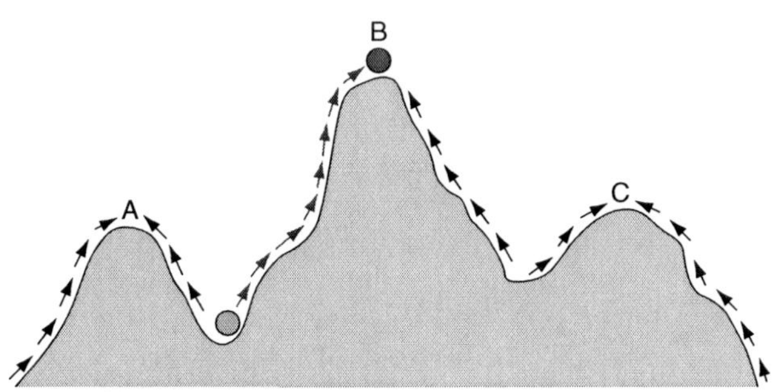

Figure 3.1 A rugged fitness landscape

problems are complex, 'i.e. made up of many non-linearly interacting components, then local incremental search... generally leads to highly suboptimal and path-dependent' solutions. (See Section 4.6 on path-dependence and Section 8.3 for a mechanism which addresses at least this difficulty via a collectivization of enquiry. See, generally, Section 7.1 for a thorough discussion of such complexity and its implications for effective divisions of labor in the cognitive sphere.)

3.1.7 The burdens of judgment

The political philosopher John Rawls provides a nice phrase for and a good synthesis of the kinds of problems we have been considering in this section. The phrase is 'the burdens of judgment' and the synthesis is as follows:[19]

> [T]he burdens of judgment... are the many hazards involved in the correct and conscientious exercise of our powers of reason and judgment in the ordinary course of our political [or cognitive] life... [and they include:]
>
> a. The evidence—empirical and scientific—bearing on the case is conflicting and complex, and thus hard to assess and evaluate.
> b. Even where we agree fully about the kinds of considerations that are relevant, we may disagree about their weight, and so arrive at different judgments.
> c. To some extent all our concepts... are vague and subject to hard cases; and this indeterminacy means that we just rely on judgment and interpretation... within some range (not sharply specifiable) where reasonable persons may differ.
> d. To some extent (how great we cannot tell), the way we assess evidence and weight... values is shaped by our total experience, our whole course of life up to now; and our total experiences must always differ....
> e. Often there are different kinds of normative considerations of different force on both sides of an issue, and it is difficult to make an overall assessment.
> f. Finally... any system of social institutions is limited in the values it can admit so that some selection must be made from the full range of... values that might be realized.... In being forced to select... we face great difficulties in setting priorities and making adjustments. Many hard decisions may seem to have no clear answer.

All these factors—and others such as fundamental uncertainty—provide a context in which it may make sense to look toward a collectivization of enquiry, so that these factors might be ameliorated in some way.

3.1.8 Implications and provisos

One implication of this analysis of boundedness is, of course, that all our decisions are made on the basis of inadequate models of the situations that are relevant to them. If descriptions of these situations are inexhaustible, then it follows that the particular *and finite* descriptions we actually use to conceptualize these situations are unavoidably incomplete. Similarly, as Marengo *et al.* make clear, for complex problem-spaces. In the case of fundamental uncertainty or of contestability, the issue is not, perhaps, incompleteness in a strict sense. The resulting models are nevertheless inadequate. Insofar as they are specific on some point of concern, they ignore either the uncertainty attendant on that specificity or the contestation of the description.

A second implication is more directly relevant to our enquiries on this occasion. As Brian Loasby has pointed out,[20] '[t]hough all models are incomplete, they are incomplete in different ways; and different kinds of incompleteness confer differential advantages and differential weaknesses.' He continues: 'The reason for having a variety of models for complex problems is not to increase the chance of finding one that is a good fit, but to provide a group of models with different fits, and which between them provide a good prospect of finding a good solution.' *In other words, it is precisely boundedness and its implications which make it sensible to think collectively about the processes and outcomes of knowledge production and, especially, to look, in hope, to a collective approach to these processes.* To put this in terms of the 'fitness landscape' discourse (see Section 3.1.6), finding a more than merely local optimum is more likely when two conditions obtain: (1) we have a number of different enquirers exploring the fitness landscape from different starting-points and (2) they are able to communicate with one another. (See Section 7.5.) In this case, while one enquirer might mistakenly conclude that A is a global maximum, from the fact that, as he continues to the right, from A, overall value decreases, a collective to which he belongs might well discover the superiority of B—e.g. because another enquirer, working from a different starting-point, discovers it as her local maximum and, crucially (and this is where issues about social comparison become important), because he and she are able to communicate and through communication to establish the superiority of B to A.

Perhaps the most important reason, of all there might be, for socializing epistemology[21] is that, once we understand that it is the community, rather than the individual, which makes and evaluates knowledge, we are in a position to understand how individuals working together in a community setting can be more efficient and effective makers and evaluators of knowledge than individuals conceived of as working alone and isolated could possibly be. In a certain sense, this is an obvious point. Nevertheless, it has taken nearly 20 years, since the publication of Steve Fuller's *Social Epistemology*, for its force to be widely understood and explicitly recognized within the mainstream philosophical community.[22]

Of course, we need to understand, as I am sure Loasby and others *do* understand, that, given the factors we have identified, a collective approach to enquiry is still *not* an approach which *overcomes* the deep infirmities of our condition and, if we take my analysis of contestability seriously, the incompleteness of our evaluative apparatus. Such an approach may, however, *ameliorate* these infirmities and that incompleteness—it gives us, potentially, more information, more calculating power, and so on, even though it does not give us—nor could it give us—unlimited information and deliberative skills. In the face of inexhaustibility, complexity, contestability, and the like, merely multiplying the powers of finite beings does not radically transform the situation, and bearing this in mind will be crucial to understanding both why it is and when it cannot really be helpful to collectivize enquiry. (This is one of the reasons for my cartographic, rather than a strictly normative approach. (See Section 9.6.) Where there is no optimization without remainder at *this* level, there is no single path that is the best path for a community of enquiry to take. Here, as with first-order problems, we will look for a diversity of approaches.)

Notwithstanding this qualification, and getting down to particulars, Kerr *et al.*, as we have seen, provide a useful summary of some of the key mechanisms of social facilitation of knowledge production—of the mechanisms, in other words, that ameliorate our fallibility and finitude. Let us consider them and, later, add others from related work. (See also, and perhaps especially, Section 7.2.)

3.2 Replication

Consider two individuals, A and B, who are working on a problem, P, within a particular paradigmatic framework. If A simply duplicates B's endeavors in relation to P, there might be some slight advantage over the situation in which just one of them works on P. If they both work in the same way on the same problem, then it is more likely, for example, that errors will be discovered. (Kerr *et al.*[23] refer to this as 'fellow members catching and correcting one's errors.') If A draws one conclusion about some aspect of P or of its solution and B draws some incompatible conclusion, then, if each is aware of this discrepancy, they may realize that something has gone wrong and try to discover what it is and how to correct it. (The assumption that enquirers *will* recognize that something is awry simply because there is a disparity of judgments is not *philosophically* innocent; after all, there are situations where more than one right answer might be available—that, certainly, is what some relativists think.[24]) This mechanism is unavailable, unless she repeats her own procedures, to either A or B if they are working alone or in isolation from one another. (And, it might be said, even repeating her own procedures may not, in view of 'the private language argument,' be enough to ensure effective 'replication.') This is, presumably, the point of the practice of replicating scientific experiments, or of getting a second opinion in the case of high-risk medical or legal problems.[25] (I suppose it hardly needs adding, here and in the sections which follow, that the ability of replication and other such mechanisms to deliver an 'assembly bonus' depend on the ability of replicators to share the results of their enquiries with other members of the community, which depends, in turn, on both parties overcoming whatever inhibitory factors might also be in play.)

3.3 Error reduction

Kerr *et al.* refer to 'random error reduction simply though increasing the number of unbiased judgments being integrated.'[26] This constitutes an allusion to a substantial body of empirical materials and to one of the most famous results of analytical social science, namely Condorcet's Jury Theorem. Sunstein provides a useful summary:[27]

> To see how the Jury Theorem works, suppose that people are answering the same question with two possible answers, one false and one true. Assume, too, that the probability that each voter will answer correctly exceeds 50%. The Jury Theorem says that the probability of a correct answer by a majority of the group increases toward 100 percent as the size of the group increases.

Of course, the various assumptions on which the Jury Theorem rests and the various restrictions to which it is subject are not unproblematic—why assume that *all* voters are more likely than not to be correct? Why restrict the choices they consider to two? On the other hand, the result has been generalized[28] and the basic idea is easy enough to grasp—the errors of some members of the community of enquiry may be 'cancelled out' by the judgments of others.[29]

3.4 Decomposition

More significant assembly bonus effects, however, depend on a more thorough-going division of labor.[30] Suppose that P can be decomposed into sub-problems p_1 and p_2. One sub-problem can then be assigned to A and one to B and, working *in parallel*, they may be able to solve the problem $P = (p_1 + p_2)$ more quickly and effectively. Once again, one of them working alone might choose to work this way—i.e. divide the problem into sub-problems and then work, *serially*, on the sub-problems, putting together their solutions to construct a solution to the problem *per se*. But, in this case, while we might get the same effect, we get it over a longer period of time, other things being equal. A working serially is less efficient in this sense than are A and B working in parallel. (It is an interesting issue how this point interacts with the more specific 'logic of patches' which we consider in Section 7.6.)

As we will see (Section 7.3), a division of labor is not a panacea, even when the 'facilitating conditions' can be engineered and the 'consistent pattern' of resistance to cooperation and collaboration overcome. In particular, there are design issues that mean that considerable coordinative work has to be done to ensure that divisions of labor respect coherence requirements—that what A does in one realm fits or combines appropriately with what B does in another realm. We need to ensure, in other words, that P really can be decomposed (without remainder or overlap) into p_1 and p_2. And, even when coherence is achieved, we may find that it is achieved at the cost of a certain arbitrariness—that we do not, in dividing labor coherently, necessarily divide it optimally. This deserves and will receive a much fuller discussion later in Chapter 7.

Of course, even these very elementary observations point to a potential trade-off. While it may be better in one way to allocate A and B separate tasks (as in the decomposition of P), it will be worse in another way, as they cannot then cross-check each other's work in the same way as when they are working in the replicative modality. Of course, if we add C, A's replicator, and D, B's, to the mix, then we can combine the benefits of replication and decomposition. Doing so, however, may create other difficulties, particularly in relation to some of the social comparison pressures which I have mentioned and to which I will return.

To summarize, and as Fischoff and Johnson, put it:[31] 'Having a colleague can alleviate some cognitive difficulties. For example, dividing information-processing responsibilities can reduce information overload; having someone else review one's work may catch some mistakes.'

3.5 Interpretation

Another possibility doesn't depend on the decomposition of the common problem, but, rather, on its interpretation.

Perhaps A understands P in a certain kind of way P_A and B in a different way P_B. Maybe these differences reflect different senses of the context in which the problem should be embedded, or different interpretations of the common paradigm in terms of which A and B address P, or, indeed, different balancings of the conservative and innovative dispositions. (I have argued elsewhere that the paradigm should be treated as an *interpretandum*, not an *interpretatum*.[32]) In this case, A will neither duplicate B's work nor focus on one element of a common problem (while B focuses on another). Rather, A and B will pursue two different solution pathways in relation to this common problem.[33]

Of course, their divergent work must be anchored in something common, perhaps standards for problem-solutions, if this approach is to improve their efficiency and effectiveness. Nevertheless, it is easy to see how this might work. P_B leads to a solution of the common problem which, by their common standards, is recognized as being better than the solution (if any) which P_A leads to. (I have argued elsewhere that the standards don't need to be common for this to happen. Perhaps B's solution is *dominant* over A's—that is, better by *any* standards, including different standards, that A and B might have.[34]) Here, the parallel processing of A and B enables them to explore more of the domain of enquiry than in either of the first two examples (of replication and decomposition). (I revisit dominance in Section 6.1.3.)

Once again, we can, and do, combine all three of these mechanisms. According to P_A, P can be decomposed into $p_1{}^a$ and $p_2{}^a$, whereas, according to P_B, P must be decomposed into $p_1{}^b$ and $p_2{}^b$. If we allocate teams, rather than individuals, to each of these sub-problems, then we can reap the benefits of all three of the mechanisms I've identified so far—replication, decomposition, and interpretation. (That there can be difficulties even when we decompose problems in multiple ways is a startling finding which I consider in Section 7.3.2.)

3.6 Expectancy disconfirmation and divergent thought

The points I've discussed so far are really, if you like, about the *additivity* of collective enquiry. What one person does can be added to what another person does to get an outcome that neither of them could have gotten without the other's contribution. The next mechanism is, if you will and to switch analogies, about the *dynamics* of collective enquiry. It is a mechanism which, when engaged, enhances the interaction between individuals and delivers, potentially, outcomes which are more than the sum of what the individuals' separate contributions would have been if they had not been in interaction with one another but, rather, merely putting together the information or ideas which each of them had on her own.

The mechanism now to be considered in fact facilitates something more and deeper, something that goes beyond the mechanical aggregation of individual packages of data. What it facilitates is precisely the habits of mind and of discourse that make some genuine synthesis of diverse points of view and information possible. Kerr *et al.* put the point this way: '[T]he voicing of alternative positions during group discussion may produce *expectancy disconfirmation*, which has been shown to undermine judgmental confidence and promote more systematic processing of information.'[35] It is easy to see why, if we are interested in the efficient and effective production and evaluation of knowledge, we might want to promote more systematic processing of information. But how does *undermining* judgmental confidence facilitate this? One mechanism might be this.

We will shortly come to see (see Section 4.2) why, but of course it is 'obvious' from an 'intuitive' point of view, that an individual, A for example, might adopt a positive test strategy in relation to his own favored interpretation (Section 4.2) of a problem or paradigm. He will, in other words, look for those packages of information which support or at least are compatible with his own preferred thinking on the topic of discussion. And this may be reasonable enough given the boundedness of A's rationality as an individual enquirer. He can't consider *all* the data or even all the various relations between data and paradigm that might be considered by a less bounded being than he is. But what if B, because she foresees some personal advantage in dissenting (see Section 5.4), is sufficiently disinhibited to give A information which does not confirm, but rather, stands in some tension with his favored interpretation P_A? In this case, A may be provoked to consider why it is that B says what she says. As De Dreu and West, drawing on Moscovici,[36] put it:[37]

> [M]inority dissent is surprising and leads majority members to wonder why the minority thinks the way it does. Rather than seeking verification and justification of the minority position, majority members seek understanding of the minority position to better reject it... [T]he tension produced by minority dissent and the majority's desire to resolve this tension produce divergent thinking: Majority members consider the issues from multiple perspectives, one of which is suggested by the minority.

De Dreu and West report the results of experiments, in effect; they are telling us what people confronted by dissenting opinions do actually (sometimes anyway) do. Christopher McMahon, approaching the matter philosophically and hence not purely empirically, arrives, however, at much the same analysis. He says:[38]

> To offer an argument concerning what the evidence supports is to offer it for criticism by others. And in criticizing it, others regard it as manifesting malfunction. Thus, to the extent that others present cogent rebuttals, the results will be the purging from the individual offering the argument—and from the group, on the assumption that the process is public—of one form of malfunction. For completeness, it should be mentioned that we can also imagine a process of collective reasoning that is explicitly aimed at eliminating malfunction, with the construction of a shared pool of reasons as a by-product. Here, the members would take turns, each offering his reasoning to the group for criticism and receiving from the others the benefit of the reduction or elimination of malfunction. To the extent that all observe this process of serial criticism, the result will be shared pool of criticized arguments.

Whether in the De Dreu/West or in the McMahon formulation (and they are not exactly the same), this is precisely the kind of engagement with one another that permits not only the assembly of different packages of information (and hence a knowledge bonus); it is, as well, a form of engagement which permits genuine creativity, or, as Cass Sunstein puts it,[39] a form of 'synergy.'[40] Of course, as I've already indicated, we cannot assume that dissent from a majority point of view will be treated respectfully enough to provoke divergent processes of thinking and hence improve problem solving. It follows from the fact of dissent that a *rational* agent will be anxious to understand the basis and strength of the dissenting opinion. It cannot be assumed, however, that every *empirical* agent will be anxious and hence engaged in this way. Whether she will be depends, in fact, on two matters: (1) How she understands the nature of the disagreement and (2) how she understands her own responsibility to get things right. I consider the former matter in Section 5.6 and the latter in Section 5.8.

3.7 Robustness

One of the greatest benefits of collaborative epistemic arrangements is, of course, the ways in which what one person does can confirm or validate what others are doing.

In a useful recent contribution, James Bohman gives an interesting account of 'robustness' or, as he puts it, 'the ways in which conclusions are strengthened in light of evidence obtained through relatively independent techniques and theories.'[41] Whereas many discussants have focused on the importance of unbiased participants in securing robustness of this kind, Robert Goodin has provided a novel and persuasive account of the way in which we can learn from, and have our judgments altered by the contributions of, putatively biased observers. He thereby adds a new layer of sophistication to the idea of expectancy disconfirmation.

In particular, Goodin distinguishes between contributions which, as he puts it, go either for or against the grain of the observer's known biases.[42] Suppose that a scientific study purports to identify the presence in tobacco smoke of certain carcinogenic agents. Suppose that we are broadly skeptical of claims of this kind. If the study were carried out by scientists already known to be 'anti-smoking' and were sponsored, say, by the Cancer Council or some similar organization, the findings wouldn't necessarily engage our interest or shift us in our antecedent position. Suppose, though, that the study was supported by tobacco industry funding and carried out by an 'in-house' scientific team? This might well shake us. Goodin provides a good account of the underlying dynamics:[43]

> We all know that there are a great many more-or-less equally legitimate ways of setting up our experiments, and a great many experimental results we might more-or-less equally legitimately report

> deriving from those different set-ups. By the same token, there are a great many auxiliary hypotheses that we might reject in order to protect our core hypotheses from empirical disconfirmation. Combine all that with the hypothesized disposition of the putatively 'biased' observer to choose among them to report the result among those that is most consonant with her bias. If someone who is looking for any way she can to report a result that favors the tobacco industry, and finds none (reporting instead that 'tobacco smoke is carcinogenic'), then that surely counts as strong evidence that it is.
>
> Given what we know (or confidently suppose we know) about *that* researcher's biases, if she tells us that tobacco smoke is carcinogenic, we can be pretty sure that she has explored all the alternative ways of setting up and interpreting the experiments, and she has been forced kicking and screaming to that conclusion. Coming from her, it really *must* be true. Or anyway, in light of what we know about her predispositions and the 'extra steps' that those lead her to take to check out the results that she is now reporting 'against her grain,' we attach a very much higher conditional probability to that report being true, given she was the one reporting it, than we would to the same sort of report coming from someone from whom that sort of conclusion is 'only to be expected.'

This is, if you will, a particularly salient form of expectancy disconfirmation and the conclusion which we draw when this sort of (admittedly unlikely) event occurs is more robust, other things being equal, than conclusions drawn merely on the basis of 'concurrent' judgments—i.e. judgments which concur or agree with what we might already independently have decided about a matter.

3.8 Two kinds of assembly bonuses

There are two ways of understanding the 'assembly bonus' which might be delivered by a collectivized process of enquiry.

In the first case, the bonus associated with assembling what different individuals know and value might be a collective judgment (a judgment endorsed by each of them) about the matter they're enquiring about—and a judgment which is better in some sense than the judgments (and they may well be plural) that they could have arrived at working in a more individualistic way. A and B put together what they know about some problem P and, as a consequence, they both become convinced that P is to be understood and solved in some specific way which they both accept as being the best way.

In the second case, the assembly bonus is not an improved *collective* judgment, but, rather, an improved basis on which individuals can form their own judgments about the matter they're enquiring about. In this case, A and B assemble what they know and value and, as a result, they collectively deliver an improved 'pool of reasons,' as Christopher McMahon puts it,[44] which each of A and B can draw on to develop, on this improved basis, their own (quite possibly different) judgments about the matter they're enquiring about.

This distinction is relevant, I believe, in understanding why collectivization of enquiry might be important even when convergence of judgment is not easily obtained or, perhaps, even properly sought—as, for instance, where there is contestability (see Section 3.1.5) or fundamental uncertainty (see Section 3.1.2). Even where we cannot reasonably expect that A and B will come to agree, we can reasonably hope that their non-convergent, or at least not necessarily convergent, courses of deliberation might be improved by the collectivization of enquiry, and this might well be the case where that collectivization leads to the improvement of the quality and perhaps even quantity of whatever it is that contributes to the deliberative process.

Obviously, the improvement of the 'pool of reasons' is relevant both in situations which do aim at convergence of judgment (e.g. settled science) and in those which do not (e.g. aesthetic judgments). Nevertheless, the value of an improved 'pool of reasons' may well be different in these different situations. In the one, we value the improvement of the pool as a facilitator for the convergence of judgments. In the other, we value the improvement of the pool as a facilitator of better judgments even in the absence of convergence. In both cases, of course, as McMahon puts it,[45] '[e]ach [enquirer] benefits from the common pool of reasons by being able to make a judgment about a particular matter that is more responsive to all the relevant reasons, including the reasons provided by the fact that others have made certain judgments, than it would otherwise have been.'

3.9 Interim summary

Notice that all the mechanisms discussed in this chapter require the various individuals to assume specific roles and to relate to one another in certain kinds of ways—especially, that they *share* the information they have obtained while pursuing their separate enquiries and that their subsequent, collective deliberations about this information be conducted in certain ways. Crudely, we get a bonus only if we assemble the information which each has separately gathered. Garold Stasser put the key issues very clearly:[46]

> Although there seemingly is merit to the idea that groups can, in principle, benefit from pooling members' knowledge and expertise, the degree to which they, in practice, realize this benefit is open to question. The presumed benefits depend on two conditions. First, at least some of the members must bring to the group unique information that others do not have; otherwise, there is no 'unique' information to pool. Second, such unique information must be mentioned during discussion; otherwise, the collective decision does not benefit.

In earlier work, I argued that it is irrational not to share such information, specifically, that refusing or failing to do so manifests one of a number of forms of epistemological fundamentalism, such as dogmatism or relativism.[47] Perhaps this conclusion was too strong and perhaps, as a pluralist,[48] I ought to have seen that even in 1990. After all, even an enquirer is not solely an enquirer. She, like all of us, has a diversity of interests and values and, even if my argument were enough to establish that enquirers have an *epistemically decisive* reason-to-share (information), that argument could not, in itself, establish the conclusion that such a reason, generally, ought, as a matter of individual rationality, always prevail over such other, potentially countervailing reasons which people also have.

In other words, if, like Kitcher,[49] we want to know 'what are good ways for communities to organize their efforts if they want to promote the collective acquisition of epistemically significant truths,' we will need to consider, not just what it is epistemically rational to do, but, more importantly perhaps, and certainly more challengingly, how to identify circumstances and motives which facilitate the realization of enquirers' potential for epistemic rationality (as far as this is possible) and, hence, for reaping the bonus associated with the assembly of knowledge obtained by rationally conducted parallel processing or expectancy disconfirmation. We have before us, to repeat Bourdieu's formulation,[50] the question:

> What are the social conditions which must be fulfilled in order for a social play of forces to be set up in which the true idea is endowed

> with strength because those who have a share in it have an interest in truth, instead of having, as in other games, the truth which suits their interests?

I approach the task set by this question indirectly. First, I consider, in Chapter 4 immediately below, the consistent pattern, identified by Hinsz *et al.*, which can inhibit the rational interaction of individuals. Then, in Chapters 5–8, I will identify a number of motivational, institutional, and environmental factors which may work to counteract these inhibitions and thus permit the assembly bonus to be reaped.

4
The Consistent Pattern

In Chapter 3, I considered how individuals might, in a community of enquiry, play distinctive roles—associated with the ideas of replication, decomposition, and interpretation. In any of these ways, we get, if you will, more diverse (potential) input to a process of collective deliberation about the common problem the community is engaged in addressing. But we will realize the assembly bonus only if these individuals are able, at some point, to put their distinctive information together so that it can play a role in their deliberations which enhances those deliberations.

Another idea from psychology may make it clearer what is involved in 'the combination of different pieces of information that are separately inadequate to produce accurate judgment but which together make a new, emergent solution possible.' This is the idea, which we owe to Stasser, of a hidden profile. As Stasser puts it:[1] 'A hidden profile exists when the superiority of one decision alternative over others is masked because each member is aware of only part of its supporting information, but the group, by pooling its information, can reveal to all the superior option.'

Let's see, schematically, how this might work. Stasser and Birchmeier provide a good account:[2]

> [D]ecision [option] X has four items of supporting information (denoted X_1, X_2, X_3, and X_4) and Y has three items of supporting information (denoted Y_1, Y_2, and Y_3). If the items of information are equally important and relevant to the decision, X is the better choice. However, [if] the supporting information for Y is commonly known, whereas the supporting information for X is mostly unique [distributed across rather than shared by the individual members of the group] ... [then], before discussion, each of the three members will

> tend to favour Y based on what they individually know because they do not see the bigger picture. If [A, B, and C] ... mention their unique information during discussion, they will see that X is the superior choice and endorse X even though they all favored Y before discussion. That is, the group discussion can reveal the hidden profile of information that favors X.

So long, in other words, as A, B, and C remain ignorant of what the others know and what the significance for their common problem is of what the others know, there is a *hidden* profile with respect to that problem: a good solution is latent in something that, between them, they know, but it is hidden because it has not yet become common knowledge for them. If, however, these individuals are so situated and so inclined that they *can* put their uniquely held information together, then they will reap a bonus in the form of a solution to the problem they were working on that relies on assembling the distinctive information which each of them had separately developed.

What the terminology here—'bonus' and 'hidden' in particular—helps us see is that collective problem-solving is only more efficient and effective than individual problem-solving when certain kinds of social and attitudinal conditions exist. If A, B, and C can't or won't share their information, then assembly will not occur and no bonus will be realized; the solution-relevant profile of knowledge will *remain* hidden. Perhaps these individuals are reluctant to share because they are competitors, not collaborators. Perhaps they find it hard to uncover the hidden profile because of dysfunctional 'sampling dynamics,' as Stasser puts it;[3] they do not, for whatever reason, sample, in their conversations, all the (problem-relevant) information that, individually, they have.[4]

In this chapter, I want to identify 11 mechanisms which psychologists, economists, management scientists, and other investigators have identified which might stand in the way of uncovering a hidden profile. (These mechanisms by no means exhaust those which might be mentioned.) Later, in Chapters 5–8, I will consider some ways in which these inhibiting mechanisms might be overcome. This book, if you like, is about the 'essential tensions' between the inhibitory mechanisms which I discuss in this chapter, and the disinhibitors which I discuss later. In any event, and to anticipate, the eleven mechanisms are:

- Social comparison;
- Heuristics;
- Paradigms and their effects on deliberation;

- Norms of joint agency;
- Persuasive arguments;
- Path dependency;
- Equilibrium traps;
- Institutionalization of dominance;
- 'Stickiness' of information;
- Principal/agency issues; and
- Affect and affect contagion.

4.1 Social comparison

A has some piece of data, d_A, relevant to the problem that the community is addressing; B has another relevant piece of information, d_B. Putting the two pieces of information together enables A and B to solve the problem P. But putting the pieces of information together requires both A and B to *say* what information they have, and it cannot be guaranteed that they will do so. Indeed, in many settings, both experimental and naturalistic ones, either or both of A and B may not share with the other, or with the larger group to which both belong, the distinctive information which they have (and which is relevant to the problem they are working on). This may seem irrational (and it often is), but it is very commonly observed. (Notice, by the way, that their reluctance to share information affects all the mechanisms we have considered, including, especially, replication, interpretation, and expectancy disconfirmation; the impact of such reluctance is by no means confined to the mechanism of devolution, which explicitly requires the amalgamation of distinctive 'pieces' of a puzzle.)

Why wouldn't A and/or B share such information? One reason is that A, say, isn't interested only in using his interactions with B as occasions for sharing problem-relevant information which he may have and she may lack. Indeed, in many situations, this may not be his primary motivation; A is likely always or at least very commonly to be interested (also or maybe even predominantly) in using his interactions with B and others to establish or enhance his standing in and with the group and its other members. (This is, if you will, a concrete example of the distinction I made earlier (Section 3.9), between the epistemically rational course of action and the course of action which may seem best when other non-epistemic interests are also in play.) Two distinct mechanisms seem to be at work.

First of all, people concerned about fitting in will tend to 'oversample shared information ... and undersample in their discussion information which one of them has and the others do not have.'[5] (This is the dysfunctional 'sampling dynamics' referred to earlier.) Schematically, rather than A mentioning the unique information d_A (which he has and B doesn't) and B mentioning her unique information d_B, A and B are more likely to mention and even discuss some item of data, call it d*, that is common knowledge between them, in the sense that both have this information and each knows (or at least believes/guesses/assumes) that the other has it too. In this way, A can see that he and B are similar and B can see that she and A are similar, and, in certain, common, situations, this is an important (non-epistemic, but still salient) interest for both of them.

Secondly, people concerned about their social standing are likely actually to change their opinions so that *antecedent* diversity (which might have been tapped to produce an assembly bonus) is actually reduced. As Cass Sunstein puts it:[6] '[P]eople want to be perceived favorably by other group members, and also to perceive themselves favorably. Once they hear what others believe, they adjust their positions in the direction of the dominant position.' Or, as De Dreu and West, put it:[7] 'There is good evidence that being a dissenter within one's team can negatively influence one's affective well-being. Being a dissenter is stressful . . . and may increase anxiety and fear of being expelled from the group,' and may, accordingly, induce a dissenter to change his opinion to conform to that which is normal (and perhaps even normative) for his group. (See Section 4.11 for more on the affective side of communities of enquiry.)

The diversity of information holdings that antecedently characterizes A and B, in our schematic example, may, in other words, be attenuated by this process of adjustment. Perhaps B's information is relatively easily assimilated to the dominant position. She may provide this information. But if A's information is less easily assimilated to this position he (a) may not provide it and (b) may indeed no longer himself give it credence (and thus have nothing distinctive to express). He adjusts his own position, thus eliminating one of the elements of the hidden profile and thus reducing the prospects for an assembly bonus. As Kiesler and Cummings put it:[8] 'The presence of others seems to increase a person's concern with what others think and increase involvement with the group and the group's activity. . . . The presence of others increases conformity through its effects on felt surveillance and social pressure.'

Two provisos should be registered here and will be relevant more generally.

(1) While it is important not to pontificate in the abstract, it would appear, in these schematic cases, that withholding information and changing one's views are epistemically irrational. After all, they (the withholding and the changing) don't depend on rational argumentation showing, for example, that there is no benefit in expressing the information or, indeed, that the data-claims are mistaken. The (epistemic) rationality of these responses is not the point, however. Should these phenomena be reasonably common, they need to be taken account of if we are to produce practical, rather than utopian recommendations about or empirically grounded analyses of collective knowledge formation. If they are not taken account of, as they typically aren't in normative theorizing, then we cannot be certain that epistemically appropriate norms will be taken up and, accordingly, we cannot be sure that they will deliver the beneficent effects whose delivery constitutes their primary rationale. We therefore can't be sure, in fact, that the norms are indeed appropriate. (See Section 2.3.)

(2) While there may well be institutional or other arrangements, within scientific communities for instance, which lessen or perhaps even mostly or entirely eliminate the deleterious effects of these two mechanisms: (1) we will benefit from identifying these arrangements explicitly and (2) we cannot assume that even well-regulated epistemic communities are entirely immunized (institutionally or otherwise) from these effects. (These same provisos will apply to the other mechanisms which I identify.)

4.2 Heuristics

While A and B may well seek out information held by the other, they will not and, on account of bounded rationality (Sections 2.3 and 3.1), cannot canvas *all* the information which each of them holds. In this sense, a sampling or interrogative strategy or heuristic approach is vital and unavoidable.[9] (See Section 6.2.2.) The necessity of a sampling strategy, however, does not mean that the strategy typically adopted by enquirers is an optimal or even a good one. Indeed, there is reason to believe that the typical sampling strategy, again in a variety of settings, is not well adapted to achieving the desired assembly bonus. In particular, as Suls points out:[10] '[P]eople generally use a positive test strategy, that is, seek hypothesis-consistent information.... Routine use of a positive test strategy may discourage people from actively seeking out or appreciating the opinions of dissimilar others when it could have great utility.' (Again, it will be apparent; perhaps especially in this case, that such a limitation on sharing will have adverse effects on the mechanism of expectancy disconfirmation, as well as the other, perhaps more 'mechanical' mechanisms.)

In our schematic example, if A already has information d_A, then his interrogative strategy in relation to B is likely to focus on the elicitation from her of information bearing on the credence and solution-relevance of d_A. Accordingly, he is unlikely, unless d_A is evidentially linked with d_B, to elicit, through his positive test strategy, precisely the information which might, when combined with his, deliver the assembly bonus. (This is one, perhaps a novel, way of appreciating the Popperian emphasis on falsification.[11] An institutional norm favoring disconfirmatory data may sustain sampling heuristics that better facilitate the achievement of assembly bonuses. I return to this point in Section 5.4.)

Indeed, we can see how a positive test strategy might work in tandem with social comparison mechanisms to strengthen the inhibition of information sharing. In our schematic example, A and B are more concerned with fitting in and with getting confirmation of their own information than they are with using their diversity to leverage an assembly bonus. A and B meet and, inevitably, one of them will begin the conversation, let it be B. (It is not a trivial matter that the conversation between A and B will proceed serially; one of the key lessons of this entire work is that temporality matters, and, in particular, that it matters who goes first. See Section 4.6.) What will happen? Actually, there are several possibilities; here are two.

B sees A as typical of the majority in the community to which both belong. In this case, her interest in fitting in may well dictate deference to A and, if A too sees himself as typical in this sense, he may well, using the positive test strategy, seize the conversational initiative and focus their discussion of the issue on confirming his own information, d_A. In this case, there is no assembly bonus because B never introduces her information d_B.

B does not feel constrained in retaining conversational initiative and so defaults to a positive test strategy in relation to her information d_B. Insofar as A considers B typical of their community, he will defer to B's agenda and, again, there is no assembly bonus because A never introduces his information d_A.

Of course, this is highly schematic, and I want to insist, in Chapter 6, that there are indeed non-inhibiting ways of ensuring that people feel that they fit in—i.e. ways of satisfying this interest that do not inhibit conversation promoting the realization of the assembly bonus.

Nevertheless, these two scenarios, however schematic, give us a feeling for the issue. As long as individual epistemic agents are only boundedly rational, specifically in also being influenced in their interactive styles by affiliation interests and not solely by epistemic interests, then we can expect some inhibition of the facilitative possibilities on which the assembly bonus depends. Learning how or to what extent an epistemic community might have engineered or evolved these inhibitions out of their activities will therefore be a vital concern.

4.3 Paradigms and their effects on deliberation

Heuristics are one mechanism for coping with inexhaustibility (see Section 3.1.1). Another mechanism is provided by what Thomas Kuhn called the *paradigm* for research activity, which, *inter alia,* directs attention both toward and away from aspects of an inexhaustibly rich field of perception or cognition. As Kuhn puts it:[12]

> [S]omething like a paradigm is a prerequisite to perception itself. What a man sees depends both upon what he looks at and also upon what his previous visual-conceptual experience has taught him to see. In the absence of such training there can only be, in William James's phrase, 'a bloomin' buzzin' confusion.'

Whatever the undoubted utility of the enquirers' commitment to a paradigm of research, however, it can make them more resistant to assimilating and using information or ideas which they might encounter in their conversations with uncommitted (or even less committed) colleagues (who might, of course, feel equally committed to some other point of view). One way of characterizing these kinds of commitments is by saying that the enquirers become 'principled' in their approach to the topic of enquiry; they have principles to which they are committed more firmly than they are to other elements which might be relevant to their investigative tasks and which they might indeed encounter.[13] And this makes them resistant to genuine dialogue with others who do not share their specific commitment(s). As Carnevale and Pruitt put it,[14] 'each party digs in its heels to defend its own interpretation.' They continue: '[P]arties who prominently mentioned principles tended to be hostile and rigid and to fail to reach agreement.' More importantly, such discussants fail even to take in what their conversational partners have said. Where these partners have, for example, a datum which, combined with other data already collected, makes available a potential 'assembly bonus,' this bonus will not in fact be realized for the actual combination of data will not occur so long as it is inhibited by the principled commitments of the various parties.

In fact, the availability of a paradigm, and enquirers' principled commitment to it, has a tendency to *channel* research and to raise barriers to new ideas. David Teece provides a good account, in the context of the management of large organizations. Echoing some ideas from another sphere of enquiry which I will sketch in Sections 4.6 and 9.2, he says:[15]

> Technology often evolves in certain path-dependent ways, contoured and channelled by what might be thought of as technological paradigms (Dosi, 1982). A technological paradigm is a pattern of solutions to selected technical problems which derives from certain engineering relationships. A paradigm identifies the problems that have to be solved and the way to inquire about them; within a paradigm, research efforts become channelled along certain trajectories. Relatedly, new product and process developments for a particular organization are likely to lie in the technological neighbourhood of previous successes.

It is, as we already noted in Section 4.1, easier for enquirers to engage, conversationally, on terms set by an already available paradigm than it is for them to make contributions which have to be developed *ab initio* without the benefit of 'common knowledge' or taken-for-granted assumptions among their interlocutors. While this conservatism permits the elaboration of a research program, it also inhibits the sharing of information that is 'outside the square.' (We consider another, rather different, idea about paradigms at Section 6.2.1.)

4.4 Norms of joint agency

Drawing on the widely cited work of Michael Bratman,[16] Deborah Tollefsen provides a very telling analysis that shows an irony at the heart of cooperative activity—namely, that precisely the features which sustain it also may limit it and, especially, may stifle the free discussion of a diverse range of views. (The same might well be said of the paradigm and its role in enquiry.)

As Tollefsen points out, what Bratman calls 'shared cooperative activity' 'is exhibited in small research teams such as those found in microbiology, cognitive psychology, and neuroscience, among others,' As she puts it,[17] such teams will be held together by '[s]hared values, intentions, and commitments [which] will be experienced as "ours."' As she notes, it is precisely '[t]he stability of shared cooperative activity [that] makes it possible for groups to progress and achieve long-term goals without having to rethink every action and sub-plan.' The price of such cohesion, stability, and hence productiveness, however, is not insignificant. As Tollefsen puts it:[18]

> Given the normative and intentional structure of teamwork, it is clear why there will be pressures to suppress dissent within the context of a team's deliberation. In addition to all of the biases or decision vectors at play in the deliberation of a team, there are these additional constraints brought about by the very nature of the agency in which they are engaged. Raising alternative viewpoints or challenging the team's findings will cause the group to rethink its shared intentions, subplans, and shared values. It may cause participants to question the dissenter's commitment to mutual support and their commitment to the joint activity or project. Such a rethinking has the potential to produce a deterioration of the group's ability to function as a unit.

Suppression of dissent, or, more commonly, self-censorship of unshared ideas or insights, is functional in one way and dysfunctional in another. It is functional insofar as it enables the conditions of shared cooperative activity, or joint agency, to be maintained... and insofar as such activity or agency is successful in dealing with the issues or problems the groups is engaged with. It is dysfunctional insofar as it results in the suppression of precisely those contributions by team members that might provide an assembly bonus.

Here we are close, surely, to the essential tension which Kuhn identified (see Section 2.1) between the productivity of collaboration within an agreed framework and the downside risk that is associated, precisely, with that productive enhancement of individual, unguided effort—namely, the loss of flexibility, both individually and collectively, that can result and that can be a threat, particularly when the agreed framework no longer provides for a productive engagement with the domain of enquiry.

4.5 Persuasive arguments

One explanation for why people interacting in a group might not express the full range of their problem-relevant knowledge is, as we've seen, that they might not want to be seen to be too different, in various respects, from their peers. This may be troubling, personally and emotionally, but it can also, as Tollefsen has reminded us, destabilize institutional or cultural arrangements which have productive benefits. Another explanation for the suppression, including the self-suppression, of diversity draws on broad social comparison mechanisms, but does so in a way that gives more prominence to the argumentation that we might expect to find in a scientific community or other communities of enquiry. This explanation is styled 'persuasive arguments theory' and Forsyth provides a good overview:[19]

> [P]ersuasive arguments theory... assumes that individuals base their decisions on the number and persuasiveness of the arguments that they have for and against each position.... [G]roups can, in most cases, generate more arguments that support the position endorsed by the majority of the group or the position that is most consistent with dominant social values, in part because *members may be more willing to express arguments that are consistent with social norms.*

The underlying mechanism is simple enough, though the final clause in the above quotation (to which I have added emphasis) alerts us to some complexities.

Suppose that there are twice as many group members with A's general approach to some problem than there are with B's. Assuming that there is some intra-group diversity within these two cohorts (a point I will return to in Section 6.1.1), it follows that A's cohort should be able to generate more information of a kind compatible with A's approach than B's cohort will be able to generate in support of B's approach. As Sunstein put it:[20] 'The key is the existence of a limited argument pool, one that is skewed (speaking purely descriptively) in a particular direction.' But this may well mean that, not only are the B-supportive arguments 'outnumbered' by the A-supportive arguments, but, as with the self-censorship in play with generic social comparison mechanisms, that some at least of the B-supportive arguments (a) are not even expressed and hence never come into the count and (b) are factored out of the thinking even of the individuals who were originally aware of them. In this case, B's divergent approach gets absorbed, if you like, in the majority approach and, for practical purposes, might as well not have been present in the first place. As Forsyth says, group 'members may be more willing to express arguments that are consistent with social norms,' and, accordingly, less willing to express arguments that are not in accordance with such norms.

4.6 Path dependency

Another reason why enquirers might not consider the full range of 'dispersed knowledge' can be explicated by reference to the idea of *path dependency*. The basic mechanism is simple enough.[21]

As I've already indicated, it is not a trivial matter that discussions are temporally extended and *ordered*. In particular, someone speaks first and then others join in. But the first speaker has what Randall Collins calls[22] 'first-mover advantages' in the sense that other individuals wanting to participate in the discussion will find it easier to respond to what she, the first-mover, has said than to start an entirely new thread of conversation. There is therefore a tendency for subsequent contributions to the discussion to be oriented to this emerging *line* of discussion and, accordingly, for discussion to be confined to a narrower range than might be theoretically available. (Remember: there are (see Section 3.1.1) inexhaustibly many things to talk about.)

Where the discussion arrives is, in other words, dependent on the path on which the first speaker set out. As Xueguang Zhou puts it,[23] '[e]xperience and information gained in the past decrease the cost of learning in the neighbourhood of the familiar area and put a higher price tag on explorations into unfamiliar territories.' There are, in other words, and continuing with the economic terminology, increasing returns on the investments that those enquirers have made that have formed the basis of discussion. Because they have established a discursive center of gravity, it will be easier for other individuals joining the conversation to orient to that center (than it would be for them to start off in a different direction). As Susan Leigh Star puts it,[24] '[t]his results in what philosopher William C. Wimsatt calls "generative entrenchment"...That is, decisions made early in the life of a project ramify throughout development, in much the same way that small changes early in the life of an embryo have crucial downstream effects.' More concretely and more narrowly, but still pertinently given our larger interests, as Steven Woolgar puts it,[25] 'the scientific laboratory comprises the materializations of earlier scientific decisions and selections. The knowledge products which stabilize and solidify are embodied in technologies *that set the scene for the next set of interpretive decisions* (representations).'

Of course, first-mover advantages and generative entrenchment are not automatic consequences of the temporality, if you will, of collective deliberation. Christopher McMahon, for instance, seems to argue that they can be overcome, and the point is not a bad one. He says:[26]

> And the most that can be hoped for in the way of constructing a pool of reasons is that all the relevant reasons known to anyone in the group will be presented. But—especially when discussion ends when no one has anything more to say—it can be expected that the natural flow of discussion will achieve this result. What others have already said may prompt people to thoughts that they would not otherwise have had. So the manner in which the pool is constructed will depend on the flow of discussion. *But* at least if time pressures are not severe, and thus people who have already spoken can speak again if something that is said prompts new thoughts, *the different possible flows should all result in (approximately) the same pool.*

Notwithstanding his seeming confidence on, and the seeming reasonableness of, this point, McMahon himself qualifies his statement when he says, pointing the way toward conclusions closer to those drawn and defended here, that his conclusions depend on the assumption, surely contrary to fact, especially given the analysis presented here in Chapter 4 and already sketched in Section 2.2, that 'everyone who wants to speak has the opportunity to do so, and can say all that he or she thinks worthy of being said.' He continues, and this might be a motto for my own work on this occasion: 'This may be unrealistic unless there is some active management of the process.' His own account of that 'management' is defective, in my view, on two grounds—its psychological naïveté and its relatively crude understanding of the possibilities.

Notwithstanding these qualifications, what all this points to is that some knowledge is, if you will, more equal than others (with apologies to George Orwell). Different members of a community of enquiry will have different opportunities to have their distinctive knowledge registered with the community as a whole. If what A knows doesn't cohere with a conversation which has already become generatively entrenched, then, on account of generic social comparison mechanisms, he may not even offer this knowledge as a contribution to this conversation. If what B knows does cohere, then she will feel freer and easier about adding her mite to the accumulating mass of information and, precisely by adding it, she will contribute to the further entrenchment of this conversational direction, thus making it even less probable that a contribution like A's will be made, or, if made, heard.

Notice, especially, the importance of Zhou's and Woolgar's contributions. Each of them identifies a mechanism which is not purely social-psychological (as with generic social comparison pressures). What Zhou points out is that, when one person (or group of people) does a lot of work pursuing a particular line of enquiry, they create, at the same time, at least typically, a *method* for enquiry. And this means, of course, that new entrants to the field (see Section 6.7) may find it easier to work in that field if they simply adopt the already existing method, thus further entrenching its dominance. Similarly with Woolgar's point. New entrants to any field inherit the *apparatus*, both physical and institutional, that earlier entrants have developed. It will always be easier, whatever other possibilities might exist in the abstract, to use *this* apparatus than to try to develop, *ab initio* and *de novo*, apparatus of your own.

But all this means, of course, that potential diversity is reduced and, with it, opportunities to explore more thoroughly the field of enquiry. Well-entrenched existing practices exert, if you like, a centripetal force on enquirers. (Both methods and apparatus, and of course paradigms and heuristics are part of the institutional machinery of 'joint agency' (see Section 4.4).)

4.7 Equilibrium traps

The kind of 'generative entrenchment' that results from 'first-mover advantage' may actually be compounded by another effect which has been reported in the management literature, under the heading of 'equilibrium traps.' The basic ideas are simple enough.

Suppose that the 'first mover' proposes a particular approach to a problem. There are two possibilities: (a) the proposed approach is a sensible one and (b) the proposed approach is not a particularly good one. (I have tried to avoid terminology which suggests in too specific a way just how good (or bad) the first mover's proposal actually is.) You might think that in the second case, where the first mover's proposal was not actually a very good one, *second* movers would have an opportunity to shift the discussion away from its starting point and to introduce their own ideas or information. To a certain extent, this will depend on the position of the first mover in the local social hierarchy. Obviously, if her position is a prominent and respected one, then, when she makes her initial move, others, to ensure that they are not too disadvantaged competitively, will be likely to 'track' that move (to cover themselves) and hence to deprive themselves of precisely the opportunity to seize the initiative that this mistake on the part of the first mover objectively opened up to them. (This is not inevitable of course and my analysis doesn't depend on its being inevitable.) As Zhou puts it:[27]

> If the [first mover's] decision is a miscalculation, ... imitative behavior by others tends to cause them to stumble as well, helping the risk taker to maintain its relative advantage in the social hierarchy ... [and] leading to what Akerlof (1984) has called 'a lower level equilibrium trap' (p. 32). The resulting inefficiency at the collective level conceals the initial misstep by the leader.

Approaching matters from a more sociological perspective, Paul DiMaggio and Walter Powell describe a similar mechanism, which they label 'mimesis.' As they put it:[28] 'Uncertainty is also a powerful force that encourages imitation. When organizational technologies are poorly understood (March and Olsen, 1976), when goals are ambiguous, or when the environment creates symbolic uncertainty, organizations may model themselves on other organizations,' especially those that are perceived as being successful or leaders in the particular field of endeavor. Accordingly, even when a perceived leader is mistaken in its approach to the problem or issue, it is likely that others will follow its lead rather than strike out in different, and possibly more productive, directions.

The seeming stubbornness of such imitative behavior is only compounded, of course, by the 'loss aversion' which follows upon the 'investment' which imitators (and the first mover herself) make in the positions which they take. Having committed themselves to a particular point of view, having invested in its cogency, they are more likely to be averse to abandoning it, seeing this as a loss, rather than an opportunity for gain or even for comparative advantage. As Robinson and Kray put it,[29]

> [b]ecause the seller in a transaction begins with an endowment, this person may be expected to be especially loss averse. In decision making, this kind of loss aversion results in sellers demanding more money to part with an item than they would be willing to pay for it, had they not already been endowed with it ....

Or, in other words, once an enquirer has committed herself to a particular point of view, she will demand more compelling evidence to abandon that point of view than would have been required to reject it if she had, for instance, been in a state of epistemic indifference. And so, having been led, by imitation of the first mover's mistaken 'innovation,' to adopt a particular position, she will, even when alternatives to that position become more attractive than they were initially (or, indeed, more attractive in some objective sense that her current position), be relatively immobile in relation to those alternatives. She will find herself, in Akerlof's terminology, in a trap at a lower epistemic level than she might have been able to attain in the abstract. (But she never decides in the abstract, but, rather, always in precisely such concrete settings as we have here sketched. On this matter, see also Section 9.2.)

4.8 Institutionalization of dominance

I have already mentioned that 'first mover advantages' may be substantial and may be difficult to overcome, even when the first mover has moved in a not particularly promising direction (as compared with some other directions in which he might have moved). Of course, the 'first' in the expression 'first mover' may sometimes have a hierarchical rather than a strictly temporal interpretation. Sometimes, the person whose movement is first of any significance in defining a course of enquiry is not the person who makes the temporally first movement, but, rather, the person who, because of already accumulated credibility, needs to be attended to by others in the community. Steve Fuller put it this way:[30]

> This brings us to what Robert Merton (1973: 439–59), in another of his classic euphemisms, called the 'principle of cumulative advantage'.... Merton originally had in mind the selective advantage afforded to scientists trained at elite institutions, considered in relation to their ability to accumulate grants, publications, posts and honours.... As in capitalism, so too in science: the rich get richer, the poor get poorer.

In other words, cumulative prestige, if you will, can make an individual first among equals in certain respects that are relevant to delineating the subsequent course of enquiry. If we add to this the points, already made, about generic social comparisons factors, as well, of course, as the specific issues associated with path dependency and equilibrium traps, we encounter a rather gloomy overall picture of the prospects for enquiry to identify even local optima.

A, because of the prestigious position from which he enunciates his point of view, possesses first-mover advantages even when he is not, temporally speaking, the first mover in a particular discursive context. B, because of A's prestige, and because of generic social comparison processes, will be loathe, even if she disagrees with A's point of view, to explicitly challenge A's proposals or ideas. Indeed, she may gradually come to adopt this point of view as her own. Despite the possibly suboptimal character (in some objective 'fitness landscape'; see Section 7.1.2) of what A has said, others, and not just B, will find it easier to join A than to oppose him and will gradually add to the stock of persuasive arguments favoring A's position, thus leading to an equilibrium of discussion in this field of study that may well be at a lower level (relative to a 'bird's-eye-view' objective assessment) than might have been attainable had other information been added or other forms of reasoning been canvassed.

Closing the circle that was drawn around A's original 'inheritance'—e.g. his attendance at a prestigious graduate school, his sponsorship by sectoral 'big guns'—A and his cohort will, in due course, capture the social apparatus which enables them (and others of their ilk) to reproduce and extend the reach and the recognition of their ideas. They will, for example, put in place a methodology and an instrumentation that others will naturally tend to use because it will be easier to do so than to start afresh. (See Section 4.6.) As Pierre Bourdieu puts it:[31]

> In the forefront stands the education system, the only institution capable of securing the permanence and consecration of official science by inculcating it systematically (the scientific habitus) upon all legitimate recipients of educative action, and in particular, upon all new entrants to the actual field of production. In addition to the institutions specifically charged with consecration (academies, prizes, etc.), the established scientific order also includes the instruments of circulation, in particular the scientific journals which, by selecting their articles in terms of the dominant criteria, consecrate productions faithful to the principles of official science, and exercise de facto censorship of heretical productions...

4.9 'Stickiness' of information

We owe to the management theorist Eric von Hippel the lovely coinage 'sticky information.' It points to another source of inhibition in information sharing.

According to Hippel,[32] the stickiness of a piece of information is defined in terms of the 'expenditure required to transfer that unit of information to a specified locus in a form usable by a given information seeker' at that locus. Hippel goes on to identify several reasons why information transfer might not be cost-free, from which it follows, we might speculate, that such transfer might in fact be inhibited, thus reducing the prospects for an assembly bonus to be delivered through interactions between diverse members of the community.

One of Hippel's reasons for information stickiness is especially pertinent to communities of enquiry and, in particular, to communities which exhibit a high degree of intra-community specialization of roles, as is certainly characteristic of the scientific community. (On the division of labor, see Chapter 7.) As he puts it,[33] '[i]nformation stickiness can . . . be high because organizations must typically have or acquire *related* information and skills to be able to use the new knowledge that may be transferred to them.' Hippel offers the example of an artist working with a programmer, with the intention of developing software to facilitate the computer-assisted generation of works of art. Each knows something relevant to the project that the other does not know, but what each knows is, in some cases at least, so embedded in the specific skills and cultures of their different professional communities that it will be difficult for them to establish a surface of contact that will enable them to realize an assembly bonus. Crudely, they inhabit different worlds, to borrow a Kuhnianism, and, unless they have some prior experience of working across the disciplinary boundaries by which they are separated, their interactions and negotiations will be protracted, fumbling, and costly, and the profile of knowledge bearing on their common problem may well remain hidden. Even more crudely, they don't know how to talk with one another; they lack a common idiom; the knowledge of one is incommensurable with that of the other, and vice versa.

Putting the matter schematically, A, the artist, has some information d_A about the aesthetic side of their common problem and B, the programmer, has some information d_B about software routines and computer interfaces, but these two packets of information are expressed in such disparate languages and are associated with such divergent modes of thought and professional practice that it will not be enough, to produce an assembly bonus, merely that A and B overcome their social comparison anxieties and express what they know. They will also need to find or create, on the spot, an idiom in which these packages of information can be expressed so that they *can* be assembled. But, of course, until such a translation has been effected, the stickiness of these packages of information is an inhibition to their assembly and, hence, to the realization of the potential bonus.

(Information stickiness thus represents a limit, if you will, on the scope of norms of 'joint agency' (Section 4.4). Perhaps rather ironically, it blocks information transfer precisely to the extent that it preserves informational diversity. It thus functions quite differently from social comparison mechanisms, which often block transfer of unique information precisely by flattening out antecedent diversity.)

Another reason for information stickiness is related to the idea that some knowledge is tacit rather than explicit (and that tacit knowledge will typically be stickier than explicit knowledge).[34] So, for example, part of what the programmer knows is explicit enough; the difficulty with some of this knowledge is that it is not intelligible to the artist. But some of what the programmer knows (and the artist too) is not propositional and hence can't be transferred between them primarily because it cannot be translated into some mutually intelligible *lingua franca*, but must, instead, be acquired by direct practical exposure. Some of what the programmer knows is embodied in her skills and her attitudes, indeed in what we might call her 'sensibility,' and she may be as unaware of this knowledge as the artist is and, accordingly, incapable of transferring it, cost-free, to the artist. Here too an assembly bonus will accordingly be difficult to realize.

(Again, the two provisos already registered, especially the second, are pertinent. There is no reason to believe that these difficulties have proven insuperable in the case of science and other analogous activities. Indeed, there is conspicuous evidence that they have not. My project is to establish, if you like, that there are obstacles to overcome in realizing the facilitating potential of diversity (and, hence, that we should not simply take this potential as given) and, indeed, that these obstacles have been overcome in certain specific ways which it is useful to identify explicitly and in some concrete detail. This is part of a cartographic exercise, if you will—to chart the pitfalls and to explicate the ropes and tackle that we use to cross over them. See Section 9.6.)

4.10 Principal/agency issues

One of the reasons why members of a community of enquiry might not share information or skills which are dispersed amongst them is discussed, in the economics literature, under the heading 'the principal/agent problem.'[35] The key issues in this case are to do with delegation and with the diversity of individual motivations. A classic example, schematically expressed, is this. B, his supervisor, delegates to A responsibility for performing some task T. A acquires knowledge about and skills in addressing T which B, typically, will neither have nor have any reason to acquire. B will monitor and discuss with A the outcomes of his activities in relation to T, but the 'inner workings,' if you like, of these activities will be largely, and for efficiency's sake rightly, invisible to B. The invisibility of the 'inner workings' creates, however, a situation in which information about T is asymmetric (in the language of economists) and, more importantly, it creates opportunities, which he may not seize of course, for A to conceal certain aspects or indeed certain outcomes of his performances in relation to T. For example, since B lacks knowledge of the 'inner workings,' A may be able to create value in addressing T that he does not pass on to B. B is interested, *ex ante*, only in certain outcomes or 'key performance indicators,' and if these do not completely capture all potentials inherent in T (and how could they given inexhaustibility? see Section 3.1.1), and if A, in addressing T, produces other outcomes which do not, as it were, appear on B's 'radar,' then he can seize the value created by his activities for himself, despite their potential relevance to B and to the larger projects which she supervises.

Of course, A's incentive to withhold information is not necessarily presumptive. There may well be incentives to share information which override this rather narrow analysis. See, for example, Section 5.4. Nevertheless, the mechanisms identified later need to do work to counter mechanisms such as are involved in the principal/agent problem if assembly bonuses are indeed to be realized.

4.11 Affect and affect contagion

One of the ways in which boundedness in rationality shows itself is through the influence of emotions, moods, and other affective states on individuals' thought processes and on their interactions, socially, with one another. Even more so here than in the other cases already considered, I need to register a proviso. While negative affect and its contagious spread through a group may well inhibit the facilitative possibilities of diversity in relation to knowledge generation, positive affect and its spread may well enhance these facilitative possibilities. In other words, the mechanism I am going to describe in this section is not one that unequivocally inhibits the facilitative possibilities which we have already identified. (I will return, in Section 6.8, to the possibly beneficent effects of *positive* affect contagion.)

What is affect contagion, and what does it tell us about communities of enquiry? According to Grawitch and Munz,[36]

> Affect contagion occurs through unconscious mimicry processes, in which group members begin to mimic the affective expressions of other group members (e.g., facial expressions, postures, voice tone...). Due to cognitive and physiological feedback mechanisms, this mimicry leads to affective convergence within the group... Over time, these processes result in consistent levels of affect within the group setting (also known as *affective tone*...), which influences the group atmosphere...

In other words, positive or negative affect may, under certain conditions, spread through an enquiry-focused community to become a predominant tone for that group. And the establishment within a group of a predominantly negative affective tone can, according to Grawitch and Munz, have considerable impact on the effectiveness of that group as a generator of problem-solutions. They say:[37]

> Because of its destructive qualities, a negative group atmosphere does not permit group members to discuss potential problems openly, resulting in the identification of less difficult and more superficial problems.... In the brainstorming stage [of problem solving],... a negative group atmosphere leads group members to censor their ideas for fear that other members will criticize those ideas.... In the solution development stage,... a negative group atmosphere leads to destructive criticism, causing group members to censor themselves to avoid confrontation.

Or, in other words, where the affective tone of a group is predominantly negative, diversity of points of view and the expression of these points of view will be (comparatively) inhibited and, again, assembly bonuses are less likely to be realized. To what extent these results are directly or unproblematically relevant to typical communities of enquiry is not straightforward. After all, many such communities are 'virtual,' rather than 'face-to-face' and, of course, many such communities, of whatever kind, are characterized by positive affect, which can enhance information sharing. The circumstances in which we can expect to encounter positive group affect (and hence the facilitation of information sharing) is something I will consider in Section 6.8. In the meantime, my main point here is that affect is a potentially inhibiting factor that may need to be overcome or managed if assembly bonuses are to be realized.

4.12 Interim summary

In Chapter 3, I considered some of the ways in which diversity might serve enquiry. In this chapter, I have outlined some of the ways in which common social-dynamic mechanisms may inhibit this fundamental (Kuhnian) mechanism. James March provides, I think, a good summary of the implications of my analysis to this point:[38]

> In some ways, organizations are able to overcome information processing limitations, being more capable of parallel processing, of inventorying knowledge, and of mobilizing expertise... At the same time, however, organizations introduce additional problems, problems of retention of information and communication, of coordination and conflict.

Indeed. In the next three chapters, I will consider how these additional problems might be overcome. In particular, I will consider accountability, the novelty premium, asset ownership, and, most importantly perhaps, the constructive role of the fundamental attribution error. I will also consider the crucial contribution of communication, and, in a Kuhnian vein, the role of values (rather than more tightly defined norms) in mediating social interaction in communities of enquiry. I consider two important feedback mechanisms, the performance loop and the environmental loop, before reconsidering a device which I first introduced about 15 years ago, i.e. the wave model.[39]

5
Motivating Diversity

An assembly bonus may well be available for communities of enquiry (relative to what their members might produce working individually), but this will depend, *inter alia*, on how well these communities neutralize certain inhibitors of group deliberation, such as social comparison, heuristics, information stickiness, and the like. Crudely, if members of the community are constantly looking over their shoulders trying to second-guess what others think about their common problem, they will not make efficient use of their distinctive information (and skills) to produce the available bonus. In this chapter, I will consider some of the motivations which might be available, either by happenstance or by design, within a community of enquiry which might provide some counterpressure to the social comparison and other inhibitory mechanisms which we have surveyed in Chapter 4. In particular, I will consider:

- Accountability
- Benchmarking
- The republican element
- The novelty premium
- Asset ownership
- Attitudinal versus intellective discourse
- The fundamental attribution error

As none of these mechanisms is commonly mentioned in otherwise related discussions,[1] this chapter represents, I believe, a significant advance in our understanding of these matters.

5.1 Accountability

An observation of Hinsz *et al.* points the way forward:[2] 'One example of the effect of context is the accountability of group members to some audience or constituency... [Studies] found that the deliberation process of mock juries changed when they were expected to be made publicly accountable for their verdict...,' and, in particular, that accountable juries were more likely to consider a wide range of opinions and, hence, to draw on the diversity of their members, thus overcoming the inhibitions we sketched in Chapter 4.

The psychology is, I think, simple enough, and certainly applicable more generally than simply in juries. Suppose that A has a piece of information d_A, which is of some potential relevance to solving a problem P. On the one hand, social comparison (or norms of joint agency) within the particular group G in which an assembly bonus could be realized might inhibit A from expressing what he knows. On the other hand, social comparison with the group where the information actually originates—with his research team T, for example—may well motivate A to share the information with other members of G, even at some risk to his standing within *that* group. After all, if he does *not* do so, then, while he might reap the advantages of fitting in in G, he will not easily be able to account to his colleagues in T for his silence in the more general forum. (This presupposes, of course, a certain amount of 'transparency'—i.e. that T-members can learn of A's behavior in G. This will not always be the case, of course; recall the discussion, earlier (Section 4.10) of principal/agent issues.) If A anticipates this difficulty (of accounting to T-members for his silence about d_A in G), then he may not remain silent in G, and his information may contribute to group deliberation about P, possibly facilitating the realization of an assembly bonus. (And, of course, it may not. After all, others may not 'hear' what A says.)

Indeed, there is more to it than that. It is not just that A might, given his multiple accountabilities, be more willing to share information than he might otherwise be. It is, as well, that his own processes of thought may be more complex. (We here identify, in effect, a contrary mechanism to the mechanism of accommodation to majority opinion that I mentioned at Section 4.1.) As Helen Longino put it:[3]

> Membership in multiple communities may be a source of internal conflict, but it is also an epistemological resource. It permits an individual to compare standards and to assess (purported) knowledge

> produced and accepted in one community in reference to standards proper to another. It confers on individuals the ability to see things from more than one point of view or at least to understand that there could be more than one point of view about a given matter.

In effect, insofar as A understands himself as a *representative* of T in G,[4] he will be less inhibited by intra-G social comparison pressures than he would be otherwise. He is more likely to share information and he is more likely to engage in more complex forms of thought. This is one, specific form of the sociological ambivalence whose relevance to the community of enquiry Shadish *et al.* had already identified.[5] It might also be described as an individualized form of the 'expectancy disconfirmation' that we already considered in Section 3.6. Just as A's thought might be complicated by the fact of the dissent of another individual B, so too might A's thought in one role be complicated by the fact that A himself plays another role.[6]

5.2 Benchmarking

Accountability sometimes takes the 'personalized' form of accountability to, as above, a team or community. But accountability can also be 'impersonal,' as, for example, when we hold ourselves, as enquirers, to certain standards, perhaps expressed in the form of 'best practice' or 'iconic achievements.' In this case, we 'benchmark' our own activities and results against the achievements of others. Michael Hogg and Deborah Terry provide a useful introduction to this mechanism:[7]

> One way in which organizations may deliberately manipulate the intergroup social comparative context is by 'benchmarking.' An organization selects specific other organizations as a legitimate comparison set, which threatens the group's prestige. This motivates upward redefinition of organizational identity and work practices to make the group evaluatively more competitive.

This is clearly a crucial mechanism in any form of enquiry which aspires to excellence, for it is by developing standards of excellence, typically embodied as concrete achievements (at least the more illustrious of the Kuhnian paradigms), that the 'economy of esteem' (Section 5.3) is articulated and which provides a system of incentives and disincentives for individual practitioners. So, for example, the film *A Beautiful Mind* portrays the game theorist John Nash pursuing, as a young man, 'a truly original idea,' as doing so is, in his view, 'the only way I will ever distinguish myself. It is the only way I will ever matter.' This is his benchmark and, although the film does not offer guidance here, Nash (as portrayed) clearly benchmarks himself against, say, the achievements bound up in Special Relativity or Quantum Mechanics; they are the 'comparison set' which 'motivates upward redefinition' of his own aspirations and hence opens him up, at least in principle, in the same way that expectancy disconfirmation would. (This is closely related to the 'novelty premium,' Section 5.4.)

5.3 The republican element

Drawing on the influential recent work of Philip Pettit,[8] Steve Fuller presents a 'republican' account of science (in a non party political sense) which identifies important elements of a culture of enquiry that complement some of the considerations to be introduced in Chapter 6. According to Fuller,[9] the crucial commitment of a specifically republican community of enquiry is what he calls 'the right to be wrong,' which depends,[10] in its turn, on certain material conditions, which include the beliefs, widespread and confidently held in the community,

- that people's opinions might change for the better as a result of hearing opposing opinions;
- that people need not fear the consequences of their expressed opinions on their material well-being;
- that there is a 'public good' or 'civic ideal' to which people may appeal.

Republican polities aim to ensure that all citizens are sufficiently secure in their material circumstances that they are not inhibited from speaking their minds.

These points, especially the reference to 'material circumstances' that support self-confident expression of potentially dissenting views, are important ones in providing a counterweight to social comparison and other inhibiting pressures. Insofar as we can create, within a particular community of enquiry, material circumstances like these, then, to that extent, can we make it easier for people to share the conclusions of their own investigations and deliberations with others in their community.

Certainly, some concrete communities of enquiry embody these circumstances and do so, in some instances, in conscious recognition of their importance to the effectiveness of collective enquiry. In the academic community, for example, systems of tenure and principles of 'intellectual freedom' insulate role-players from adverse consequences 'on their material well-being' of what they might say or think, even if it runs counter to prevailing views within their particular disciplinary groups. (Such mechanisms do not, of course, entirely counterbalance the 'economy of esteem'[11] which is also, perhaps ironically, a feature of republicanism. Someone who is not inhibited by fear of losing her job may still be inhibited by fear of or even anxiety about losing favor with her peers or 'betters.')

Some editorial processes, such as the reliance, in decisions for or against publication, on the recommendations arising from double-blind refereeing provide another conspicuous, and complex, example. Some of the features of these processes are worthy of comment.[12]

(1) The editor who commissions and then decides on the basis of double-blind refereeing is assuming that the anonymity of the referee protects her from intimidation or 'pay-back' if she makes adverse findings about the merits of a paper submitted to her scrutiny. The editor is also assuming that 'adverse findings' may play either or both of two roles in improving the quality of thinking in the area surveyed by the paper. First of all, 'adverse findings' may have enough merit on their face to constitute a genuine and straightforward corrective to inadequate thinking on the part of the author. Secondly, though, even when 'adverse findings' do not constitute a straightforward counterargument or even corrective, they may, as already mentioned (see Section 3.6 on expectancy disconfirmation) contribute to enhanced thinking in this area, especially by the author.

(2) The anonymity of the author enables the referee to focus on the merits of the arguments; she is less likely, in the case of authors who have already accumulated substantial cultural 'capital' (see Sections 4.8 and 6.7), simply to acquiesce in their views. Nor, in the case of more junior authors, with little cultural capital, is she likely simply to ignore their views, as of little consequence to the 'landscape' of prestige and patronage in her field. Of course, she will still find it relatively more difficult to deal 'objectively' with papers which depart in some way from highly 'entrenched' positions (see Section 4.6) than with papers which follow a well-established line of enquiry. That sort of conservative bias is inherent in any system embodying the 'essential tension.' Nevertheless, there are countervailing pressures

(e.g. 'novelty premiums'; see Section 5.4) and her own anonymity gives freer play, than she would have in other circumstances, to any latent bias in her own individual evaluative profile toward more innovative approaches.

(3) Reliance by the editor on the findings of a process of double-blind refereeing is, in fact, his way of expressing two core commitments of the republican ethos. First of all, he thereby expresses his belief 'that people's opinions might change for the better as a result of hearing opposing opinions,' precisely by allowing what opposing opinions are produced by the process to have some effect on the author's further deliberations, and, where he agrees with these opinions, with the contents of the journal he edits, and, therefore, with the deliberative 'center of gravity,' as he sees it, in the field reported on in that journal. Secondly, he thereby expresses his belief 'that there is a "public good" or "civic ideal" to which people may appeal,' and that this 'ideal,' embodied in the standards referees are asked to apply to submissions, represents a higher standard and an objective counterweight to first-mover advantages (Section 4.7) and institutionalized dominance (Section 4.8).

As important as these points are, they need to be supplemented by another element which is often associated with republicanism, especially in its classical and neo-classical forms—namely, the role of 'esteem,' broadly construed, in promoting people's participation in collectivized forms of enquiry. Here my guide is Christopher McMahon, who, without using the label, nonetheless articulates a clear republican mechanism for securing effective participation, and one that foreshadows the mechanism to be identified in the next section, to which it therefore provides a natural transition. (I must add, however, that, as already sketched, esteem is a mechanism that cuts both ways; it can be inhibiting as well as, as I will now show, facilitative.) In any event, McMahon says:[13]

> [T]here is also another motive that can play a role in underwriting collective reasoning. This additional motive derives from the desire to maintain a certain sort of status in a group of which one is a member.... The desire to be recognized as cognitively competent will lead the members of a group to defend the views they initially express and to attack opposing views. And in doing so, each will contribute to the creation of the public good associated with collective reasoning: a pool of criticized arguments.

One way of demonstrating competence, particularly in modern cognitive contexts, is through saying something new; another way is to 'critique' a position which has been submitted for scrutiny or discussion. And these are definitely specifically republican elements, even if not ones canvassed by Fuller, in view of the analysis, for instance, by Brennan and Pettit.[14] This brings us, then, to...

5.4 The novelty premium

Having a responsibility to others, as with multiple accountabilities, is one disinhibitor. Neutralizing disincentives to share information is another. As Sunstein put it:[15]

> *Better outcomes can be expected from any system that creates incentives for individuals to reveal information to the group.* How can those incentives be provided? One possibility is to remove disincentives to reveal information, by assuring people that they will not be hurt or punished for doing so.

In fact, there are, within many knowledge-intensive communities, such as that of science, novelty premiums, as I will call them,[16] which provide precisely the needed counterweight to otherwise significant social comparison pressures. This is, of course, itself hardly a novel observation. After all, Robert Merton long ago identified novelty or originality as an important feature of specifically scientific communities... one which explained, he thought, the form and tenor of 'priority disputes.' As John Ziman puts it,[17] 'there is no real mystery about the bitter conflicts over priority that sometimes arise; the recognition of originality is at stake... [and] [o]ne's job [as a scientist] is to produce original published work.' (Novelty premiums provide an incentive to reveal unshared information, just as, on Friedrich Hayek's account,[18] price signals function to tap individuals' local knowledge, e.g. of substitution factors of production. See also Section 8.1.)

In particular, suppose that A *is* inhibited to some degree in sharing d_A in a group G, because or to the extent that he fears that he will be out of step with other members of G in believing or being committed to the truth of d_A. That is by no means the end of the story. After all, in many communities of enquiry, there are weighty intrinsic and extrinsic rewards for, precisely, providing the information that is needed for problem solving. Whatever the social pressures to conform to the general outlines of the paradigm of research in a given area, there are, balancing these, demands for originality and novelty and, hence, rewards for providing new and unexpected information which contributes to addressing the key problems for this group of enquirers. (Consider, as an example, the requirement, for the award of a PhD or for publication in a reputable journal, that the author have made a significant original contribution to the discipline.)

In this regard, I have already mentioned (Section 3.6) the Popperian emphasis on falsification, which works precisely against the so-called positive test strategy, but also, more importantly in the present context, works in favor of the provision of novel information. Insofar as members of a community of enquiry feel duty-bound to provide disconfirmatory information or are rewarded for doing so (e.g. with publication opportunities, citations of their work, etc.), then they *will* introduce information which, together with what others know, may produce an assembly bonus.

Sociologically, this is an entirely trivial observation: there are rewards for novelty in science (and other communities of enquiry). In view of the inhibitions on knowledge sharing that in fact interfere with the effective conduct of many group-based enquiries, the existence of such rewards is by no means trivial, however, to the functioning of successful communities of enquiry; without these incentives, assembly bonuses might not be delivered as efficiently and routinely as they are. (And, of course, as 'Yes Men' phenomena illustrate humorously, novelty premiums are not, if you will, an inescapable feature of the social landscape. They need, to facilitate assembly bonuses, to be built into the cultural machinery of the communities they might serve.)

Accountability, a republican culture, and novelty premiums are three mechanisms which counterbalance the social comparison mechanisms which, unchecked, have the potential to inhibit realization of assembly bonuses in group enquiry. But these mechanisms are not free-standing, and depend, in fact, on some other, perhaps more basic psychological mechanisms, which I will consider in Sections 5.5, 5.6, and 5.8. I can then, in Sections 8.1 and 8.3, consider some further, 'derived' mechanisms for the realization of assembly bonuses.

5.5 Asset ownership

In Section 4.10, I considered the threat, to the realization of available assembly bonuses, of principal/agency issues. I noted, in particular, the possibility, because of information asymmetries, that an agent would not be motivated to share with his principal *all* the fruits of the tasks which she had delegated to him. There are, of course, various ways in which this tendency might be overcome, but, as economists (being economists!) have observed, one way of promoting the more efficient and effective use of information which has potential value, collectively, is, precisely, to allocate rights of 'ownership' in this information, so that, for example, the agent becomes, in effect, a partner with rather than a subordinate to his principal and, since he owns those products of his labor which are 'excess' to those he is specifically required to produce, he can see some point in sharing these products—e.g. and most obviously by offering them for sale to his principal. ('Sale' might be a metaphorical expression in some epistemic contexts.) As Nicolai Foss puts it:[19]

> The importance of asset ownership derives from the fact that the willingness of an agent to undertake a non-contractible investment (say, exertion of effort or investment of human capital), which is specific to the asset, depends on who owns the asset.

The reasoning is not obscure. If there is, whether metaphorically or literally, a 'trade' in information and/or skills, then there is, potentially, an incentive to share these skills and/or information which may well provide a counterweight to those social comparison (and other) pressures which militate against or at least decrease the probability and effectiveness of sharing. And, indeed, we do have mechanisms like this in communities of enquiry. Even without the more rigorous and more literalist approach to ownership of ideas that is being pursued in contemporary settings by the proponents of various intellectual property regimes,[20] there are various senses in which individual enquirers (and groups) take ownership of their ideas and skills and derive further benefits from this ownership. For example, ideas which prove fecund for others working in a particular community of enquiry will come to be associated with the name of the person who has played a conspicuous role in their development or dissemination. We use the word 'Kuhnian' for example, or write of Kuhn's ideas about scientific revolution or normal science. This kind of symbolic recognition of 'ownership' (in some extended sense) is a powerful motivation to disseminate ideas when there are not specific role obligations to do so.

Or consider the ways in which an innovator (or important disseminator) can 'trade on' her association, in the minds of her colleagues, with particular ideas or findings or skills. She can, for example, hope to turn this association to account (the commercial argot is unavoidable and highly appropriate) e.g. through enhanced opportunities to

- deliver keynote addresses at international conferences and receive travel funding and even honoraria for doing so,
- recruit post-graduate students,
- win competitive grants to support her research,
- publish by invitation (rather than through the normal peer-reviewed channels),

and so on. All of these are valuable in acquiring cultural capital (see Section 6.7), and, accordingly, of managing career outcomes in a favorable way. Since these ownership benefits are accrued precisely by sharing information and/or skills, ownership, even in this purely metaphorical or extended sense, is clearly an important counterweight to social comparison and others similar pressures.

5.6 Attitudinal versus intellective discourse

As reported by Robert Rosenwein,[21] a distinction developed originally by Patrick Laughlin[22] is of some relevance in understanding the conditions under which an encounter with a divergent point of view or claim might actually lead to engagement with that claim, and, hence, expectancy disconfirmation and the more systematic processing of information. This is the distinction, in Laughlin's terminology, between intellective and attitudinal issues. As Rosenwein says,[23] '[i]n science, an attitudinal situation might hold where various hypotheses had been put forward to account for some phenomenon but where little data had been generated... [whereas] [a]n intellective situation might obtain where the accumulation of data appears to constrain the plausible hypotheses.'

This is a distinction within the realm of facts, it is important to state. In both cases, the issue is answerable to evidence and reasoning; we are not in the realm of tastes or even of values, where, ultimately, a deep pluralism might have to be acknowledged.[24] What distinguishes the kinds of factual issues which are attitudinal from those which are intellective is nothing intrinsic to the claims *per se*. It is, rather, something relational and, specifically, something about the relation of the claims to the discursive and evidential state of play. An attitudinal situation is one where the (factual) claims in play are, as yet, relatively undifferentiated with respect to strong and widely accepted evidence. An intellective situation, on the other hand, is one where the factual claims are (relatively) well settled or firmly established.

What does this have to do with expectancy disconfirmation, and other disinhibitors (working against social comparisons pressures toward conformity)? According to Rosenwein,[25] when members of a community of enquiry found themselves in an attitudinal situation, 'majority members were more likely to attend to minority hypotheses, particularly where the majority was least confident of its own hypotheses, unlike a comparison condition in which the data were more clear.' Or, in other words, when the discursive situation is relatively 'open,' before consensus has been achieved about some matter of theory, fact, or application, individual members of a community are more likely themselves to be open (to each other) and hence their discussions are likely to be characterized by expectancy disconfirmation, enhanced depth of deliberation, and so on.

Of course, we must not conclude, on the basis of Kuhn's famous distinction between normal and revolutionary science, that a community of enquiry will encounter attitudinal situations (and hence be open to minority hypotheses) only in revolutionary settings. (More generally, see Sections 6.1 and 6.2.) This represents much too crude an interpretation of Kuhn's distinction. In fact, since a paradigm is a concrete achievement,[26] it requires both interpretation and *articulation* to the also concrete (and new) circumstances in which it is progressively developed during periods of normal science.[27] Since there are, inevitably, multiple ways in which a concrete achievement might be interpreted and multiple ways in which it might be articulated to new cases,[28] at least at the leading edge, even the business of normal science will be largely played out in attitudinal, rather than intellective situations. (Indeed, Kuhn's own account[29] may support the idea that it's not until we get to the stage of writing *ex post facto* textbooks that the situation in science is settled enough to be designated intellective.) Accordingly, we can anticipate that, even in these situations, minority views will not be swamped by those of the majority, but will be given a hearing, with all that that entails. As Forsyth put it:[30] 'If a group of people find themselves in an unstructured, ambiguous situation where they have no reference point to define their expectations, perceptions, or activities, they spontaneously seek out information from others in the group.' Insofar as cutting-edge cognitive activity is conducted in such unstructured and ambiguous situations, we can count on some counterweight being felt to the social comparison pressures to which we are always subject.

5.7 Intrinsic motivation

Closely related to the idea that complex forms of enquiry are often played out in 'attitudinal' environments is the idea that, in such an environment, motivations are, ideally, 'intrinsic' rather than 'extrinsic.' So, though esteem is crucial in science, it is so at least partly because or insofar as it is linked, as Osterloh and Frey put it,[31] 'to the activity's flow..., to a self-defined goal..., or to the obligations of personal and social identities.... The ideal incentive system is in the work content itself, which must be satisfactory and fulfilling' in itself and independently of extrinsic rewards. As they insist,[32] intrinsic motivation 'is needed for tasks that require creativity.' They continue:[33]

> In contrast, extrinsically motivated persons tend to produce stereotyped repetition of what already works.... In addition, experimental research shows that the speed of learning and conceptual understanding are reduced when people are monitored. With extrinsically motivated employees, therefore, the pressure of sanctions leads to lower learning levels and the work performed is more superficial than with intrinsically motivated employees.... Second, intrinsic motivation also helps overcome the so-called multiple task problem..., where contracts cannot completely specify all relevant aspects of employee behavior and its desired outcome. Moreover, the goals to be set are often not clear to the principals.

This will bear and benefit from some explication, though the key points are easy enough to grasp.

Clearly, sustained projects of enquiry involve 'multiple task problems' in the relevant sense. We do not know, antecedently, in such situation, exactly what we are trying to accomplish. We have a problem, but, frequently, we do not have even a sketchy understanding of how it will be solved... and, indeed, in some cases, we will not even know where, precisely, the problem actually shows itself.

Of course, in these situations, a great deal of exploratory behavior will be characteristic and unavoidable. We will need to try one thing, then another. This makes it difficult for extrinsic motivations to play a crucial part in the community of enquiry. Since we don't know the dimensions of the problem, we don't have, as it were, a set of criteria against which to judge individual or team performance (and on the basis of which extrinsic rewards—e.g. esteem) could be allocated. In this case, the rewards intrinsic to the activity itself—the delight in overcoming technical difficulties (even when doing so does not lead directly to problem solutions); the interest of learning new skills (even when they cannot immediately be deployed to produce problem solutions)... all these are intrinsic motivations at least for some enquirers (and, indeed, individuals lacking access to these motives may not, by that very fact, be attracted to or easily retained in a community of enquiry engaged in genuinely creative projects).

Of course, it is one thing to say that communities of enquiry are unlikely to be staffed by 'too many' (purely) extrinsically motivated members. (If they are, they will fail, because extrinsic motivations aren't characteristically potent in attitudinal circumstances.) It is better, however, if we can understand how even extrinsically motivated individuals can come to be intrinsically motivated. Then we will have, if you will, an endogenous explanation for a crucial element of any successful community of enquiry. How this might come about is explained, I submit, by...

5.8 The fundamental attribution error

It is, I think, an implication of a broadly Kuhnian analysis that the entire community of scientific enquiry is, for broadly cultural or normative reasons, committed to an error about individual responsibility that in fact serves the interests of enquiry. Let me explain.

Suppose that A is considering some problem P and that the situation in which he considers it is marked by attitudinal uncertainties rather than intellective stability. It would be quite natural in this situation for A to excuse himself from any responsibility for not being more certain about how to address P. After all, the situation itself is unsettled and A's uncertainty simply reflects that fact, as we can easily verify when we consider, perhaps thought-experimentally, that B and others might, in the situation in question, feel exactly the uncertainty that A does. If A and B, despite other differences, both have the same experience, then, plausibly, that experience is determined (for the most part) by the situation, not the agent.

It might generally be the case that, in attitudinal situations, we attribute uncertainty, when it exists, to the situation, not the agent, but this is *not* how it is in communities of enquiry such as science. Indeed, as Kuhn himself observed:[34] 'Failure to achieve a solution discredits only the scientist and not the theory. Here... the proverb applies: "It is a poor carpenter who blames his tools." ' Shadish *et al.* provide some helpful commentary:[35]

> To help make the world predictable, humans tend to make attributions about the causes of events. Collins (1985), in fact, suggests that part of the training of new student-scientists is to teach them to make

> the attribution that the world is orderly and lawful, and to *teach them that any failures to detect such order are their own fault*. 'It is necessary for us to blame ourselves if we are to learn appropriate skills and learn to "see" the world like scientists. An impression of chaos is the alternative; science cannot be learned that way' (p. 161). Collins rightly notes that it is often difficult and frustrating to train the student to think this way, since the world is often more chaotic and recalcitrant than simple pictures of science suggest.

What Collins is describing is an extension, beyond its usual range, of what social psychologists have described as an error, indeed as the *fundamental* error when it comes to attribution (of causation or responsibility). As Cramton put it,[36] '[t]he overweighting of dispositional [i.e. personal] relative to situational determinants of behaviors and events is known as the *fundamental attribution error*.' Notwithstanding this way of putting things, the fundamental attribution error, expressed within the scientific community as a fundamental normative precept, and transmitted in myriad master–apprentice relationships, is, if you will, precisely the disinhibitor that we need to explain why, even in attitudinal situations, individual scientists orient towards one another: they will be blamed individually if they do not get things right and, accordingly, they need to consider all that might bear on their prospects for getting things right, even the ideas of others in their team or wider community.[37] As James Bohman, following, ultimately, George Herbert Mead, put it:[38]

> [D]eliberation is not just about offering and assessing our own reasons, but considering the reasons of others. This means that even as we take up our practical stance informed by our experiences and practical knowledge, we also are able to take the perspectives of others to some degree or another, when we see reasons as convincing or change our minds.

Indeed, we can be even more specific about how institutionalizing the fundamental attribution error may contribute to the progress of enquiry. As Cramton reports,[39] individuals who learn to see themselves as others see them are more likely to see themselves as responsible for their failures to solve the problems which they address. (For that is how others will interpret these failures.) But, of course, communities of enquiry such as science cultivate through various means precisely such a self-reflective attitude in their members. Anyone who has refereed other people's papers learns to read her own work with a more skeptical eye and, as she does with the strangers whose papers she assesses, to come to see the paper's failings as the author's failings, even when she is herself the author. (There are myriad other such mechanisms.)

Of course, we needn't insist that *every* member of a community of enquiry 'internalize' the 'observer's perspective' and hence attribute responsibility for failure to herself rather than the situation in which she works. As Popper in effect argued,[40] we may get a similar effect from each individual being a critic of others and being criticized by them in turn. What Popper's account does not adequately provide for, however, is 'engagement' between the actor and the observer. Why not, if I am criticized by others, simply dig in my heels and persist in my own current views? Our account answers this question. I cannot remain dogmatically fixed in my own current views if I am fearful, as I will be having internalized the critic's perspective, that these views might be in error for some reason having to do with my own incapacity.

There is, in fact, a connection to be drawn here with the materials surveyed in Section 5.3. Writing in another context, Brenda Major and Toni Schmader describe a mechanism which links cultural factors to the more psychologistic factors identified in thinking about the fundamental attribution error to explain why 'the good worker never blames her tools.' They say:[41]

> Political sociologists and psychologists have long argued that social ideologies—attitudes, beliefs, and values that are held consensually within society—often help to sustain the perception of the social system as just and fair and justify the hierarchical and unequal relationships among groups in society... Examples of these legitimizing ideologies (which Sidanius *et al.* call 'hierarchy-enhancing myths') include beliefs in a just world..., in personal causation..., in personal control..., in a meritocratic society..., and in the Protestant work ethic... These beliefs and values legitimize the status quo by holding individuals and groups responsible for their outcomes and by locating the cause of good or poor outcomes within attributes or inputs of that person or group. Indeed, the bias to locate the cause of behaviors or outcomes within individuals, rather than within situations or systems, is so pervasive in Western societies that is [*sic*] has been dubbed the 'fundamental attribution error'. (Nisbett and Ross, 1980)

That many typical communities of enquiry are meritocratic, that hard work is valued in them, and that the rewards of work and creativity are justly allocated—these are indeed some of the core beliefs of many communities of enquiry of the kinds we are familiar with. (They are, of course, the result of a great deal of 'cultural work.' I am not supposing that their existence, role, and sustenance is unproblematic or 'given.') And, as Major and Schmader make clear, it is just these sorts of beliefs that will 'frame' failures to contribute to collective projects in a way where blame or responsibility redounds to the individual contributor—and not, for example, to the intractability of the problems or issues being considered.

Of course, there are limits to what individuals will accept responsibility for. Sometimes, when enough of them have failed to advance enquiry, or when problems (Kuhnian 'puzzles') persistently fail to yield even to concerted efforts, even involving high-prestige members of the community, the circumstances may well be right for a 'revolution,' one feature of which is, of course, precisely that individuals will increasingly blame their tools and not themselves. This process was, of course, already described by Thomas Kuhn in *The Structure of Scientific Revolution*:[42]

> There are always difficulties somewhere in paradigm-nature fit; most of them are set right sooner or later... The scientist who pauses to examine every anomaly he notes will seldom get significant work done....
>
> When ... an anomaly comes to seem more than just another puzzle of normal science, the transition to crisis and to extraordinary science has begun. The anomaly itself now comes to be more generally recognized as such by the profession. More and more attention is devoted to it... If it still continues to resist... many of them may come to view its resolution as *the* subject matter of their discipline.... An even more important source of change is the divergent nature of the numerous partial solutions that concerted attention to the problem has made available. The early attacks upon the resistant problem will have followed the paradigm rules quite closely. But with continuing resistance, more and more of the attacks upon it will have involved some minor or not so minor articulation of the paradigm, no two of them quite alike, each of them partially successful, but not sufficiently so to be accepted as a paradigm by the group. Through this proliferation of divergent articulations..., the rules of normal science

become increasingly blurred. Though there is still a paradigm, few practitioners prove to be entirely agreed about what it is.

5.9 Interim summary

We are now in a position, I think, to begin to understand the field of forces which constitutes a successful community of enquiry. On the one hand, we have a potential for assembly bonuses that can be realized when individuals share information and engage with one another. On the other hand, we have a number of common social processes which inhibit precisely such sharing and engagement. But we also have a range of motivational disinhibitors which, in some situations, counteract these processes and permit the hidden profile to be revealed and the bonus to be realized from the assembly of disparate information and points of view. This is, in you will, *the empirical counterpart to Kuhn's original essential tension*. It represents, I think, a reasonable specification, in concrete terms, of a possibility already identified by Deborah Tollefsen, when she noted[43] that '[a]lthough there are factors that contribute to the suppression of dissent in group contexts, ... there are ways to mitigate the effects of groupthink.'

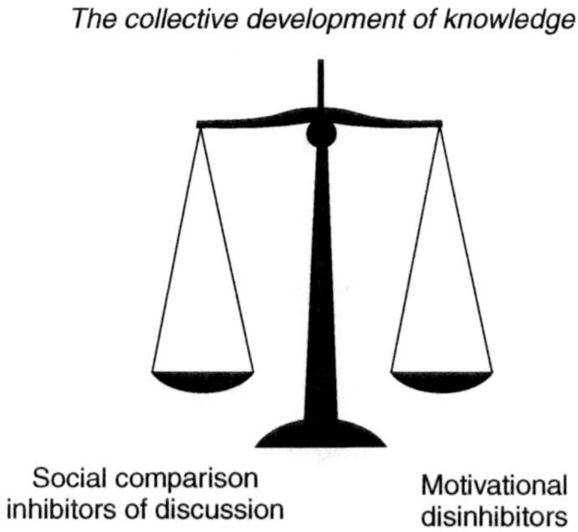

In the next chapter, I want to consider other elements in this emerging understanding of successful communities of enquiry, specifically those broadly cultural elements which may also serve to counteract social comparison inhibitors.

6
A Culture of Enquiry

So far, we've considered the community of enquiry from a normative point of view, via Kuhn, then discussed some social-psychological factors which need to be recognized, and, latterly, identified some broadly motivational facilitators, in the face of these factors, of what is, clearly, normatively appropriate. In disciplinary terms, we have canvassed philosophy, psychology, sociology, and politics as contributors to our understanding of the community of enquiry.

In this chapter, I'd like to supplement these resources by considering the community of enquiry anthropologically, in terms of its cultural elements. In particular, I'd like to consider the values in which a functioning community of enquiry will be grounded, and the role of conflict and harmony in influencing relationships and interactions among members of such a community. In the course of this discussion, I will introduce the ideas of a shallow consensus, and of a residual diversity of values interpretations. I discuss the following issues.

- Values
- Paradigms and rules
- Management by content
- Transparency
- Task conflict and relationship harmony
- Attitudes toward risk
- The performance loop

6.1 Values

I have already mentioned (Sections 3.2–6) the ways in which diversity can contribute to progress in the production and evaluation of knowledge. Insofar as A and B differ in their judgments about the facts, or the evidence, or in their interpretations of these and other cognitive elements, each can supplement or correct the other and, accordingly, both can benefit. (Because of expectancy disconfirmation, their very thought processes may also be more sophisticated.) Unstated in my earlier analyses to this effect was, of course, a point which, in relation to diversity, creates a tension, if not another essential, or anyway unavoidable, conundrum. The point is, of course, that, in some kinds of communities of enquiry, which certainly include science in many of its manifestations, the diversity of the parties is expected to compose itself, ideally, into consensus. While it can be helpful, in stabilizing cognitive claims of various kinds, to draw on a diversity of points of view, packages of evidence, and the like, such claims don't become stabilized except by becoming objects of a reasonably well equilibrated consensus. Whereas diversity with respect to some proposition is useful at one stage, its existence is, at another stage, a signal that discourse has failed, or, at least, hasn't yet succeeded. (This is the tension, if you will, between the forms of thought which Scheidel called, respectively, divergent and convergent.[1] Divergent thinking is, on Scheidel's model, associated with the creative phase of knowledge-making, and convergent thinking with the evaluative or critical phase. Both are necessary, but managing their deployment within a group can be problematic.)

This, then, is the conundrum. We value the differences between A and B but expect those differences to be resolved. How are we to understand a community in which this is possible? How are we to resolve what Star and Griesemer call,[2] in resonance with some of my own terminology (see Section 2.1), the '"central tension" in science [and other forms of enquiry] between divergent viewpoints and the need for generalizable findings'? And, in particular, how are we to understand the specifically cultural aspects of such a situation?

We can do so, I submit, via Kuhn's model of values and evaluation. (Kuhn is, on this account, a major axiological theorist.[3]) I proceed in three (main) stages. First of all, I present Kuhn's account and show that it provides for diversity. Secondly, I show how, even given this sort of diversity, it is possible to achieve consensus, to resolve differences. Finally, I show how the resolution of difference need not diminish the resources that difference provides.

6.1.1 Difference

So far, we have tended to take it for granted that different members of a community of enquiry are more than merely numerically different and, in particular, that their attitudes are different in ways which provide the facilitative possibilities which we canvassed in Chapter 2. We are now required, I think, to indicate *how* it is that individuals might be different. It will be helpful to begin with something which Kuhn himself says about the values of the scientific community:[4]

> To a greater extent than other sorts of components of the disciplinary matrix, values may be shared by men who differ in their application. Judgments of accuracy are relatively, though not entirely, stable... But judgments of simplicity, consistency, plausibility, and so on often vary greatly from individual to individual.... Even more important, in those situations where values must be applied, different values, taken alone, would often dictate different choices.... In short, though values are widely shared by scientists..., the application of values is sometimes considerably affected by the features of individual personality and biography that differentiate the members of the group.

Let me explain. Suppose that A and B are working on a problem P. A is a proponent of some theory about P, call it T_i, while B favors a rival theory T_j. Suppose that A has information d_A bearing on P, and, in particular, that d_A shows that T_i is more accurate in some sense about some matter than T_j is. (The information d_A might be the result, in this case, of a so-called crucial experiment between T_i and T_j.) It is, of course, possible that A's introduction of d_A will persuade B to accept the superiority of T_i to T_j. However, it may also happen that A and B will continue to disagree about the relative merits of these theories even if A does introduce d_A and even if B does accept the relevance of d_A to the question of theory choice. As Kuhn implies, B might continue to judge T_j superior to T_i on either of two independent grounds: (i) B doesn't understand accuracy the same way A does and, hence, doesn't think that d_A, in itself, establishes even the superior accuracy of T_i (in the sense relevant to her); and (ii) B, however she understands accuracy, considers other values important enough so that the question of *overall* choice-worthiness isn't settled by considerations of accuracy alone and hence by d_A.

6.1.2 The risk-spreading argument

This variation in individuals' application of shared values is, of course, the ground for Kuhn's 'risk-spreading argument,' which I mentioned in Section 2.1. Kuhn put the argument as follows:[5]

> [I]ndividual variability in the application of shared values may serve functions essential to science. The points at which values must be

> applied are invariably also those at which risks must be taken. Most anomalies are resolved by normal [i.e. conservative] means... If all members of a community responded to each anomaly as a source of crisis..., science would cease. If, on the other hand, no one reacted to anomalies... in high-risk [i.e. innovative] ways, there would be few or no revolutions. In matters like these the resort to shared values rather than to shared rules governing individual choice may be the community's way of distributing risk and assuring the long-term success of its enterprise.

As I've said elsewhere, this argument has been sadly neglected.[6] It is also incompletely articulated. To develop it in more detail, it will be easiest, I think, to work with Kuhn's own contrast, specifically between a rules-driven and a values-driven assessment of and commitment to paradigm work in the sciences specifically, but, really and as Kuhn's own argument makes clear, in many different kinds of communities of enquiry.

Suppose that we have a collection of individuals, each engaged in work within a particular 'disciplinary matrix.' Suppose that they are producing and assessing variants of some paradigm achievement in that discipline. While there may be some differences among the variants that these individuals produce, there cannot, on the rules model which Kuhn rejects, be any variation in the ways in which they *assess* these variants. (This may not be entirely correct; see Section 6.2.2 for a clarification.) So if A produces the variant T_i and B the variant T_j, then, if there are *rules* of assessment to whose use both A and B are committed, then, short of there being a 'tie' between T_i and T_j, either both must accept that T_i is better than T_j or both must accept that T_j is better than T_i. And, in this case, if both are rational, they both must devote their future energies to the articulation and improvement of whichever of the variants is, according to the rules of assessment, the better of these two.[7]

What's the matter with this? you may ask. It is, as Kuhn saw, risky. Why? Well, as I've tried to indicate elsewhere,[8] there are two courses of action, leading to an improvement of variants, that are not available to A and B if, because they have arrived at the same conclusion about the relative merits of T_i and T_j, they both prefer the same one of these variants.

First of all, they can't improve the better of the two variants through a *competition* between it and the worse of the two since both A and B will be working on the better variant (if they are rational).

Secondly, they can't *dual-track* the two variants with a possible longer term reversal of their original judgment—i.e. the originally inferior variant being improved to such a degree that it is now superior (according to the rules). They can't do this, in particular, since both A and B must, if they are rational, work on the better of the two variants and, hence, neither can work to improve the worse of these two.

In each case, then, A and B 'invest' all their community's resources of time, energy, and attention in the superior variant and hence lose opportunities to improve both it and its rival that they would have had if they'd been able (rationally) to work on both.

And how would a values, rather than a rules, orientation assist? Kuhn puts the matter very clearly.[9]

> When scientists must choose between competing theories, two men fully committed to the same list of criteria for theory choice [i.e. to the same values] may nevertheless reach different conclusions. Perhaps they interpret simplicity differently or have different convictions about the range of fields within which the consistency criterion must be met. Or perhaps they agree about these matters but differ about the relative weights to be accorded these or other criteria when several are deployed together.

Actually, there are, according to Kuhn, two bases on which a values approach (as opposed to a rules approach) promotes diversity in judgments... that, in turn, spreads risk and permits progressive courses of development that might not be facilitated by a rules approach.

First of all, values are not 'self-interpreting' or 'self-applying.' What A and B share, if they share a commitment to the importance of simplicity in assessing rivals in some disciplinary matrix, is not so specific and so determinative that each must, on pain of irrationality, interpret this value or criterion in exactly the same way in every situation.[10] To be sure, there may be (must be?) situations where both interpret the value and apply it in the same way.[11] But there may be situations where they don't and needn't. So A and B can look at the variants T_i and T_j in terms of their relative simplicity and, at least sometimes, A and B can reach different conclusions, consistently with both valuing 'simplicity,' about which is the simpler and, hence, the preferable variant (at least with respect to this criterion). And if they do reach different conclusions, then A might work on one variant and B on the other and, hence, they might, collectively, get the benefits of 'dual-tracking' these variants—e.g. the benefits of competition that I outlined earlier.

Secondly, however, and as Kuhn clearly implies, the values that participants use to assess variants are plural and not always perfectly 'aligned' in terms of their implications for those judgments of *overall* superiority on which a commitment is based. Let me explain.

As Kuhn points out, scientists have a number of values or criteria in mind when it comes to theory-choice. He lists simplicity, consistency, accuracy, and plausibility. As he also clearly recognizes, how a variant ranks with respect to one of these values (and relative to another variant) need not track its relative ranking with respect to others of these values. For instance, T_i might be more accurate but less simple than T_j. But, in this case, even if they agreed about all this (and, of course, they needn't, as I've already indicated), A and B might reach different conclusions about the *overall* merit of the two variants. Perhaps A thinks that accuracy is more important than simplicity and B thinks the opposite. In this case, even if A and B agree that T_i is more accurate but less simple than T_j, A might prefer T_i and B might prefer T_i. As Kuhn says, 'they agree about these matters but differ about the relative weights to be accorded these... criteria when [they] are deployed together.' And, again, we get risk-spreading diversity of judgment and, hence, the potential for the two improvement cycles which I've sketched above.

Let me summarize. Thomas Kuhn developed a 'risk-spreading argument' which showed, in effect, how scientific 'communities of practice' could achieve both solidarity among their members and diversity in the activities of these individuals. In particular, he demonstrated the value of this diversity to the enterprise in which this community was engaged. Because individuals can be united in their commitment to a value and yet divided in the ways in which they interpret and balance it against other values they are also commonly committed to, the community is able to develop variants of its paradigm achievement across a wider front than it would have access to if each individual were bound, by the canons of rationality, to reach the same conclusion about these variants as every other 'paid up' member of the community was bound to reach.[12] Diversity ensures risk-spreading and risk-spreading permits the community to explore its domain of enquiry in an efficient and effective manner.

While these points establish the possibility of difference, we must not ignore the fact that they also explain something that I have so far, probably reasonably enough, taken for granted—namely, that A and B will, at least sometimes, actually pay attention to what their colleagues say. (B attends to A's introduction of d_A because or insofar as accuracy is a value to which she too is committed: d_A is relevant to judgments of accuracy, and accuracy is relevant to judgments of choice-worthiness, and that is the kind of judgment that B is in the business of making. This is, of course, a crucial element in their 'joint agency'—see Section 4.4.) On the other hand, the crucial point here is that values, conceived as Kuhn has long conceived them, provide a basis for difference within a community of enquiry…as well as being a binding agent for that community (e.g. in directing participants' attention to one another's contributions).

6.1.3 Resolving difference

Of course, it is all very well to explain how there can be difference within a community of enquiry. (There is difference because values are, as I've put it elsewhere, *interpretanda*, not *interpretata*.[13]) We also need to explain, at least for those communities which, as I've put it before,[14] are oriented to standardization of attitudes and behavior (versus the sort of individualization that might be appropriate in other, more humanistic domains), how such differences are resolved as the community comes to stabilize claims about its domain of enquiry. To put it in an already familiar argot (Section 5.6), how, given diversity, do we turn attitudinal into intellective issues?

One answer to this question is that we do it via *dominance*, a possibility which was emphasized by Laudan and Laudan.[15] Dominance, to adapt our schematic example, works as follows. On A's interpretation or relative weighting of the value of accuracy, d_A implies that T_i is better, overall, than T_j. On B's interpretation or weighting, no such implication holds, initially. However, A might extend or enhance d_A so that, as d_A', it establishes the superiority of T_i to T_j even for B, who nevertheless (and I will return to this point) persists in her own, different, interpretation and/or weighting of accuracy. T_i *dominates* T_j in this case in the sense that, no matter how accuracy is interpreted or weighted, T_i is better, overall, than T_j for both A and B.

Another answer is that we do so, as I reported earlier,[16] via a rolling wave of dominance, in which A, having convinced B of T_i's superiority (even though A and B continue to differ about accuracy), enlists B to work on T_i and, because two heads are better than one, they, acting together, so improve T_i that C, who'd previously held out against the claim that T_i is superior, is now persuaded that it is, and, accordingly, joins A and B in a cohort of advocates for T_i. Obviously, such a process is iterative, and results in what Sunstein has called[17] a cascade and what I have called a wave of dominance.[18] (For more on the wave, see Section 6.9.)

A final possibility is that we resolve difference, in a community of enquiry, much as we typically resolve it in deliberative communities associated with, for instance, the democratic tradition—that is, by some sort of respectable social choice mechanism.

So, for example, we may have it that, among the individuals concerned, call them A, B, and C, there is no dominant option, nor is there any temporally protracted wave of consensus, but, rather, a majority point of view which, typically, will emerge as the wave itself does. So, for example, we may have it that

Table 6.1

A	$T_i > T_j$
B	$T_j > T_i$
C	$T_i > T_j$

Of course, given what I have already said about the relation between assessments along a single dimension of evaluation and overall assessments of the options *per se*, we may well have it that, although A and C agree on the overall merits of T_i and T_j, they do so on starkly different grounds.

For example, while both A and C (and perhaps even B) may agree that simplicity and predictivity are two standards relevant to the evaluation of claims (theoretical, in particular), they may disagree both about which of T_i and T_j is the simpler or the more predictive (remember that these standards are *interpretata*, not *interpretanda*), and they may differ, as well and independently, on which of these standards is, as it were, the weightier. So we might have, for instance, that

Table 6.2

	Simplicity	**Predictivity**	**Weighting**	**Overall**
A	$T_i > T_j$	$T_j > T_i$	Simp > Predic	$T_i > T_j$
B	$T_j > T_i$	$T_i > T_j$	Predic > Simp	$T_j > T_i$
C	$T_j > T_i$	$T_i > T_j$	Predic > Simp	$T_i > T_j$

or, in other words, that the agreement of A and C about the overall worthiness of T_i and T_j is underpinned by two disagreements which, as it were, 'cancel each other out'—namely about which of the options is the simpler/more predictive, and about which of these standards is the more important.

This is an interesting point to arrive at, particularly in view of the so-called discursive paradox articulated by Christian List and Philip Pettit.[19] For I am identifying as a virtue what List and Pettit say is a vice. In particular, I say that we reconcile solidarity and diversity, both of which facilitate the proper functioning of a community of enquiry, by fostering shallow agreements (in cases of 'dominance') with deeper disagreements (about the bases for these shallow agreements). (See especially Section 6.1.5.) And List and Pettit say, in effect, that this is vicious.

What they seem to require, when individuals reason about some issue to which they will have to commit themselves jointly, is that there should be at least as much agreement, if I may put it that way, about the premises as there is about the conclusion. In particular, they identify, as a one of 'two plausible demands that we might want to make on the aggregation of judgment.... that in aggregating judgments a group should reach a collective set of judgments that is itself rational.'[20] And, although the particularities of 6.1.3 (b) above differ from those of their own exemplars of collective judgment, this case does not, it seems to me, exhibit a 'collective set of judgments that is itself rational' by their standards.

For we now seem to have precisely the sort of array of 'premises' and 'conclusions' which List and Pettit think is 'irrational.' For each of two bases for comparing alternatives, we have a majority judging T_j to be superior to T_i and yet we also have a majority judging T_i to be superior to T_j *overall*. How are we to reconcile this divergence between what the 'collective' seems to think about the 'premises' feeding into their judgments of overall superiority with what the 'collective' seems to think about precisely this question of overall superiority?

As Goldman has suggested,[21] we can pose the problem in terms of a choice. Should we look at the premises or at the conclusion? If we look at the premises, we have a majority in favor of the judgment that T_j is superior to T_i in both relevant respects and, hence we might infer that the collective should conclude that T_j is superior to T_i overall, since that conclusion follows from the premises that it, the collective, collectively 'accepts.' If we look at the conclusions, though, we have a majority in favor of the judgment that T_i is superior to T_j and, hence might infer that the collective should conclude, despite the confusing situation with the premises, that T_i is indeed superior to T_j.

For List and Pettit the choice between premises-driven and conclusion-driven approaches is a tortured one. For me, it is easy. A conclusion-driven approach *preserves residual divergence* (see Section 6.1.4) and hence supports risk-spreading and, accordingly, facilitates the effective exploration of the domain of enquiry. Let me explain.

The possibility that individuals might converge on the same conclusion from different sets of premises is one that is recognized if we take a conclusion-based approach to the so-called discursive paradox. While such an approach recognizes, what the parties themselves recognize, namely that they agree about which concrete option to prefer, it does so without demanding that they should standardize on their reasons for preferring this option. It thus preserves the diversity of grounds for preference which is so important in facilitating that risk-spreading which, in turn, supports the wider *collective* exploration of the domain of enquiry.[22]

On the other hand, to demand that, because a majority of participants agrees on argumentative premises, the collective as a whole should agree on those premises (and hence on the conclusion which *they* support) seems unnecessarily Procrustean. Certainly, it will have a tendency to reduce diversity in the interpretation and application of values and will thus work against the project of risk-spreading.

6.1.4 Residual difference

So far, so good. We have an understanding of values (as mediators of decisions) that enables us to see how diversity can play a constructive role in enquiry without undermining the *telos* of enquiry (for standardizing discourses), which is, of course, to provide consensually agreed findings. The interpretability of values is the basis for diversity; the possibility of dominance or of a wave or of some 'democratic' social choice explains how consensus can nevertheless be possible. There is one further aspect of this model that has not, as far as I'm aware, been mentioned by Kuhn himself or other commentators on his work. In particular, when we have dominance, we have agreement on which option (e.g. T_i or T_j) is better while, at the same time, we retain the potential for risk-spreading diversity of assessments *in the future*. After all, the consensus *does not* (or, at least, need not) occur because (or only when) individual enquirers agree about how to interpret and balance the values they use to assess options. Even as they agree that this *option* is best (relative to whatever that evaluation is relative to), they may continue to disagree about how accuracy should be interpreted or about how accuracy and other choice-relevant values should be traded-off or balanced against one another. It's just that all *that* disagreement is irrelevant in this particular case because of the dominance of one of the alternatives or because of the particular way in which we amalgamate individual judgments to derive a social judgment.

So, while we have a consensus on the assessment of variants, giving us a sense of solidarity within the community of enquiry, we also have a *residual divergence* in values (Choo calls it 'residual equivocality'[23]) that helps support that diversity in assessments that is, in turn, necessary to spread risk and, hence, to facilitate the efficient and effective exploration of a paradigm's potentialities.

As I've put it elsewhere,[24] we have, if you like, *a perpetual motion machine of the Kuhnian kind*. We resolve differences (through dominance or social choice) but never exhaust the differences that we require to support diversity in deliberations and, hence, to spread risk (by producing more differences which need to be resolved).

6.1.5 Shallow consensus

This is, in fact, a notably sophisticated account of a *culture of enquiry*. Members of the community share values and hence orient themselves to one another—when A introduces d_A, B has reason to listen and deliberate—but what they agree on represents what I have elsewhere called a shallow consensus.[25] What A and B agree on doesn't always enforce a consensus of opinion or judgment on substantive matters in dispute between them. In part, this is because, on account of bounded rationality (see Sections 2.3 and 3.1), neither A nor B will always even realize that there is a tension in their belief systems, whether internally and individually or externally and trans-individually. They can, concretely, agree on some substantive matter of theory-choice without even realizing that they continue to disagree on the interpretation and relative importance of the various criteria that are relevant to making this choice. As Brian Loasby points out:[26]

> Fortunately, partial ignorance and bounded rationality come to our aid; for it is no easier to recognise the existence—or absence—of a consensus than of an optimum. Full consensus on objectives is not necessary to keep an organisation together; nor is it necessary to effective decision-making. The problem of integration need not be completely resolved. Even the individual need not fully reconcile his internal conflicts, since neither the interdependencies between the components of his preference function nor the interdependencies in the complex situation to which it is to be applied can be

> fully understood, or properly taken into account even if they were. Schizophrenia is indeed a problem, but moderate inconsistency is not. If this is true of the individual, then surely it must be true in the greater complexity and amid the additional obscurities of decision-making within an organisation.

It is, then, precisely the shallowness of such a consensus that sustains an essential (or 'central') tension between the demand for a convergence of judgment on the one hand and an openness to diversity on the other hand. Kuhn's model, in effect, anticipates Wenger's influential account of the role of ambiguity in sustaining a productive community of enquiry:[27]

> When combined with history, ambiguity is not an absence or a lack of meaning. Rather, it is *a condition of negotiability* and thus a condition for the very possibility of meaning... This inherent ambiguity makes processes like coordination, communication, or design on the one hand difficult, in continual need of repair, and always unpredictable; and on the other hand, dynamic, always open-ended, and generative of new meanings.... From this perspective, ambiguity is not simply an obstacle to overcome; it is an inherent condition to be put to work.

All this puts the lie, I think, to the charge, already mentioned, that Kuhn's model of science is the model of a closed society that is devoted to conformity in thought and judgment. But perhaps we have leaped from the frying pan (conformism) into the fire (irrationalism). Certainly, this is what Kuhn feared others would think about his views. He said, for instance:[28]

> [T]his characteristic of the operation of shared values has seemed a major weakness of my position. Because I insist that what scientists share is not sufficient to command uniform assent about such matters as the choice between competing theories... I am occasionally accused of glorifying subjectivity and even irrationality.

And indeed he was. See, for example, the statement by Lakatos which I quoted in Section 2.1. But Kuhn has an answer to this charge as well. As he says:[29]

> [T]hat reaction ignores two characteristics displayed by value judgments in any field. First, shared values can be important determinants

> of group behaviour even though the members of the group do not all apply them in the same way.... Second, individual variability in the application of shared values may serve functions essential to science.

The second of Kuhn's points refers, of course, to the risk-spreading mechanism which I mentioned at Section 2.1, and, as I've already indicated, this mechanism is crucial to Kuhn's status as a specifically social epistemologist and, indeed, to my task in this book. The first of Kuhn's point is similar, I think, to the idea I've several times expressed now—namely, that the sharing of values (even in a 'shallow' way) can serve to orient individuals to one another—'he's speaking my language; I'd better pay attention'.

One final point. It is interesting to note, I think, that the kind of 'variability in the application of shared values' that Kuhn refers to can be variability, though others writing about this have not always agreed, in the strictly epistemic values of the individuals involved. We do not have to posit, as some theorists have, that the variability arises because of the influence of 'sordid' (or at least non-epistemic) *motives*—e.g. the desire for recognition by one's peers. While my approach is respectably Kuhnian, it is so partly because it is explicitly pluralist about epistemic values,[30] which Philip Kitcher's account,[31] for example, seems not to be.

6.1.6 Interim summary

As I've said, this is a notably sophisticated account of the role of values in the organization of communities of enquiry. It is, perhaps, already cogently summarized by Susan Star and James Griesemer. They say:[32]

> Common myths characterize scientific cooperation as deriving from a consensus imposed by nature. But if we examine the actual work organization of scientific enterprises, we find no such consensus. Instead, we find that scientific work neither loses its internal diversity nor is consequently retarded by lack of consensus. Consensus is not necessary for cooperation nor for the successful conduct of work. This fundamental sociological finding holds in science no less than in any other kind of work.

6.2 Paradigms and rules

I have treated Kuhn as a notable axiologist, but that is not all that he has contributed to our understanding of how groups can be bound together by shared understandings that are both abstract or (ironically?) concrete enough to be objects of consensus without being so prescriptive that diversity (on which much productive collaboration depends) will be stifled in some way. In particular, Kuhn contributes two important ideas to supplement his axiology—the ideas of a paradigm and, less obviously, the idea that the methodological rules of enquiry, which might be current in some community, take the form, largely, of proscriptions, not prescriptions. Both of these ideas are invoked by Kuhn to supplement his primary explication of risk-spreading in terms of values. I consider these matters in turn.

6.2.1 Paradigms

Perhaps the most important, and often overlooked, point about the notorious (Kuhnian) paradigm is that it is, in Kuhn's words,[33] embodied in a 'concrete scientific achievement.' This, it turns out, is more important than it looks in understanding how the paradigm can be an object of consensus while supporting diversity.

Let me begin by saying that, as many others have argued and emphasized, the concrete—that which is physically embodied, 'real' in that specific sense—is, in Stuart Hampshire's phrase,[34] inexhaustibly describable; there is no a priori limit to what, even keeping strictly to the truth, we can say about it. (See Section 3.1.1.) It has parts; it has relations to other objects; all these parts and objects are themselves concrete—and all this means that there is more to be perceived, comprehended, and said about a concrete object than we can, within the limits of our own finitude, perceive, comprehend and say.[35]

Why does this matter? Well, if there is more to say than we can say, then what we do say will be selective. And if our descriptions or even silent musings on a concrete object are unavoidably selective, then, in principle, what one person says or thinks about it can differ from what other people say or think about that very same object. Each is selective and they may be selective in different ways, perhaps reflecting their other attitudes and interests, or, perhaps, the different paths (see Section 4.6) that they have traversed to reach this point. There can, in other words and in a Kuhnian patois, be diversity in the understanding of concrete objects, including the concrete achievements that constitute paradigms for scientific activity. Kuhn himself is quite explicit about this:[36]

> Scientists can agree that a Newton, Lavoisier, Maxwell, or Einstein has produced an apparently permanent solution to a group of

> outstanding problems and still disagree, sometimes without being aware of it, about the particular abstract characteristics that make those solutions permanent. They can, that is, agree in their identification of a paradigm without agreeing on, or even attempting to produce, a full interpretation or rationalization of it.

And this diversity, like that in the application of shared values, may serve functions essential to science. How could this work?

Two enquirers, A and B, agree that the concrete achievement X constitutes a model for their own activities. Because X is concrete, it has inexhaustibly many features and anyone's understanding of it will therefore be selective. Hence, when A *interprets* X (see Section 3.5), there is some prospect that he will be selective in identifying those of its characteristics which he ought to seek to model in a different way than B will be when she identifies its 'paradigmatic' characteristics. There will, in other words, be two different selective 'images' of X, X_A, and X_B. Hence, while A and B remain united in their belief that X provides a model for their own activities and while, as with shared values, that gives them a reason to talk with one another about X, they, as with the case of values, are in a position to develop this paradigmatic achievement in different ways and, thus, to spread the risk of developing the paradigm for the community to which they belong. Let me explain.

Suppose, for instance, that A's interpretation of X is based on a selection of X's characteristics that includes α and δ and that B's interpretation is based on the characteristics β and δ.[37] When it comes to exploring the relevance of X to some problem on which both A and B are working, where A looks at the potential relevance of α to this problem, B, on the contrary, look at the potential relevance of β. They thus explore a wider range of issues than they would, collectively, if both had made the same selection from the inexhaustibly many potentially relevant characteristics of the paradigm which both aim to emulate. So, again, as with values, we have diversity supporting a spreading of risk that facilitates a more efficient and effective exploration of the issues (and of the world). And, again, we also have solidarity—both A and B *are* committed to extending the reach of X, and both of them are committed to doing so subject to the discipline of those shared but diversely interpreted and balanced values whose significance to the enterprise I have already explained.[38]

Notice, furthermore, how values can mediate the sequential development of the shared paradigm. In particular, A and B will compare their two interpretations according to their (differing understandings of their) shared values and, if, say, X_A proves to be dominant, with respect to these values, over X_B, then both A and B will accept that interpretation of their shared paradigm as the favored interpretation of that paradigm—indeed it will *become* the paradigm (for them)—and will proceed thereafter on that basis—different from their starting point and, in particular, better in the degree to which it articulates the initially accepted paradigm to the issue they were originally concerned with. Of course, just as their agreement on this issue doesn't mean their agreement on the interpretation and balancing of shared values, neither does it mean their agreement on the interpretation of X_A, i.e. the newly contrived leading edge for their further explorations. For it too is a concrete achievement and hence it too is subject to interpretation—to a selection, and hence variable selections of its features. We have, again, a perpetual motion machine. We resolve differences (through dominance) but never exhaust the differences that we require to support diversity in exploratory behavior and, hence, to spread risk (by producing more differences which need to be resolved).

That values need to be interpreted and balanced (and can be differently by different people) and that concrete paradigms need to be interpreted (and can be differently by different people) are two reasons why risk can be spread in the development of scientific practice; they are two cultural facilitators, then, of the kinds of engagement between individuals that can lead to the realization of an assembly bonus in collective enquiry.

Kuhn's initial discussion of values is actually drawn by contrasting their interpretatability (and that of the paradigm) with what he saw as the overly prescriptive shaping of enquiry that might be present in a rules-based culture or community. His understanding of rules was, however, defective and the contrast, accordingly, was overdrawn. The key lies, I think, is a distinction between prescriptive and proscriptive rules, which we may now examine.

6.2.2 Rules

In fact, when it comes to balancing the importance of diversity with the demands for consensus (in some domains of enquiry), the distinction between prescriptive and proscriptive rules is highly significant. Let me explain.

Suppose that A and B are subject to the prescriptive rule R. It is of the nature of such rules that they prescribe (relatively) specific behavior—e.g. those subject to them are supposed, in a situation S to perform the specific action ϕ. Bearing in mind that rules are not necessarily, 'self-interpreting,' A and B, subject to this rule, must both ϕ in S. This is quite different from the situation where they are subject, rather, to a proscriptive rule, such as R′, which forbids the performance of some action, say ψ, in the situation S′. In this latter case, both A and B can be compliant with the rule they are subject to while performing quite different actions in the relevant situation. For it is consistent with neither of them ψ-ing that, say, A ϕs and B χs, and since performing these different actions might lead A and B in quite different 'directions,' they will be free to explore more of the relevant territory they occupy in a proscription-based regime than they would be in a prescription-based regime.[39] And this means diversity in the application of shared rules (both obey R′) and, hence, risk-spreading with all that it entails for the efficient and effective advancement of their joint cause.

This, I think, explains the attention that Imre Lakatos gave to his so-called negative heuristic, which deflected critical attention away from certain core commitments without prescribing a particular approach to the insulation of these commitments from potentially troubling empirical findings. On the other hand, Lakatos also identified a role for a 'positive heuristic,' but here too the kind of non-directiveness that we find in the paradigm and in proscriptive rules is also present, really. In particular, the positive heuristic, on Lakatos's account, is also incompletely specified and hence permits interpretation as well as fostering a shallow consensus. As he says:[40]

> Few theoretical scientists engaged in a research programme pay undue attention to 'refutations'. They have a long-term research policy which anticipates these refutations. This research policy, or order of research, is set out—in more or less detail—in the *positive heuristic* of the research programme. The negative heuristic specifies the 'hard core' of the programme which is 'irrefutable' by the methodological decision of its protagonists; the positive heuristic consists of a *partially articulated set of suggestions or hints* on how to change, develop the 'refutably variants' of the research-programme, how to modify, sophisticate, the 'refutable' protective belt.

Like the values, like the paradigm, like the proscriptive rules, even the prescriptions to which a community of enquiry must be subject are *interpretanda*, not *interpretata*... and, accordingly, they, like all the other elements, function to promote consensus (because or to the extent that they are shared) as well as sustaining diversity (which is an engine of change and ensures risk-spreading).

6.3 Management by content

Closely related to the Kuhnian point about the concreteness of the paradigm (Section 6.2.1) is an interesting discussion, by Karin Knorr Cetina, of the ways in which an orientation to the objects of enquiry can 'manage' the relationships between enquirers. She calls this 'management by content,' and describes it as follows:[41]

> The idea of management by content can be captured by two principles: it is management that maintains participants' proximity to objects or to the substance of scientific work; and management that substitutes, where possible, object-oriented structures for social authority structures.

This will take some unpacking to see its relevance to our main concerns. We can draw, perhaps, on our earlier discussion of the concreteness of the paradigm. Indeed, I can now mention something that I slighted in that earlier discussion.

Individuals working in a community of enquiry that is organized with respect to a paradigmatic achievement are oriented to the paradigm as much as they are to one another. Indeed, the paradigm could be said to mediate their relations with one another. While it is important not to hypostasize the paradigm, e.g. of Newton's Laws, of the Bohr model of the atom, any such paradigm serves as a focus of attention, of discussion, and of application for the community which is oriented toward it. We can think of it, admittedly rather extravagantly, as the *embodiment* of an idea or perspective, and, like something literally embodied, individuals can look at it and talk with each other about it, much as two shoppers, passing a display window, might look at and talk with each other about a suit of clothes that is presented there. The object mediates the interaction. Because (or insofar as) it is of interest to both (and that fact is common knowledge), they become oriented to one another. Because it has features, they have topics of discussion. Because or insofar as it has 'affordances' or 'forcing functions,' as Donald Norman calls them,[42] the object will permit or require certain kinds of behavior in relation to it, including certain kinds of joint action (see Section 4.4), as when the missile in the silo cannot be launched unless both operators more or less simultaneously turn their individual keys.

As Knorr Cetina says,[43] though without reference to these design ideas, some enquiries

> were and are marked by a constant humming of the experiment with itself, about itself. If these experiments do not need cumbersome organizational structures, it is only because of the object circuits with which they replace lines of command. It is also because in addition to these circuits they create a discourse within which the features, reactions, and requirements of technical objects are continually exhibited and expressed, and in which everyone can, in principle, assess and follow technical needs and co-shape the strategies adopted.

Of course, we can expect to find here, as elsewhere, that the objects in question are *interpretanda,* rather than *interpretata,* as I have from time to time put it. While the two shoppers may be drawn to look at and prompted to discuss the same suit of clothes, we will not always expect them to say the same things about it. Their interaction is *managed* by its content, as it were, but it is not *scripted* or *dictated* by that content. That content becomes their context, but not the entirety of their circumstances. Accordingly, as before, we can expect to find, in interesting cases of management by content, a residuum of diversity that is both a topic of and a resource for enquiry.

6.4 Interim summary

What we have established so far, in this chapter, is that communities of enquiry, insofar as they are structured, if you will, in a Kuhnian way—depending on values, proscriptions, paradigms, and management by content—provide affordances for interpretation, to adapt Norman's useful terminology. A paradigm is something you can interpret, just as a chair, to use Norman's example, is something you can sit on—that is one of the affordances of the object, in each case. If we put this together with the motivational issues we considered in Chapter 5, we begin to see, not only how interpretation (and hence the essential tension between consensus and diversity) might be possible... we begin to see, really, how such a tension might be realized, might become actual.

The reasoning is not obscure. Paradigms (or proscriptive rules or values) are *interpretanda* that creates the opportunity for interpretation. Because of multiple accountabilities, republican freedoms, novelty premiums, and the like, individual participants in a community of enquiry have motives which provoke them to take up the opportunity afforded them by the interpretability of these devices of enquiry. Indeed, these motives may well, and in a properly designed or fortuitously evolved institutional setting will, be more than potent enough to overcome the conservative or inhibitive inertia which is bound up with social comparison pressures in all their manifestations. We can (the interpret*ability*) and we will (the motivations) interact with one another collaboratively in ways which may, when conditions are right, produce the assembly bonus (while maintaining, nevertheless, the residual diversity which powers this machinery).

Several other cultural elements are also relevant however. Here are some of them.

6.5 Transparency

In Chapter 7, I will consider, in some detail, how the division of labor might work in communities of enquiry. There are, as we will see, numerous difficulties and obscurities as well, of course, as considerable potential benefits. What the very idea of a division of labor suggests, however, is something which has so far been largely implicit in my presentation—namely, that there are specialized or at least differentiated *roles* which individuals play in a community of enquiry... and, accordingly, that the ways in which they might collaborate to produce an assembly bonus (or might not) may depend, to some extent, on how roles are differentiated and what the individuals playing these roles understand about the division of labor more generally (and their roles in it).

This is a long-winded way of introducing an idea which I borrow from Carsten De Dreu and Michael West. Crudely, the idea is that, if what each individual working in a particular community does is known, at least 'locally,' to other individuals in that community—e.g. to others working on the same apparatus (as in Knorr Cetina's high energy physics teams) or in the same team, then each individual will be able, potentially, to see the ways in which what the others do conditions his own possibilities for success—they, plural, begin to see themselves *as* a team in the sense that each comes to recognize that his success depends, in part, on the contribution which others make to their joint activities. And, in this case, he is likely to become more attentive to and respectful of their contributions (insofar as these meet certain threshold conditions of competence, of course), and, accordingly, he is likely to share information with them and to expect them to share information with him. They become, in effect, 'joint agents' in the sense of our discussion at Section 4.4. Of course, as we have already seen, joint agency can be inhibitive of sharing, but it can also, especially in the presence of other motivators (the novelty premium, multiple accountabilities and the like) be facilitative. This, anyway is what Cohen and Levinthal[44] as reported by De Dreu and West argue:[45]

> Cohen and Levinthal (1990) argued that, assuming a sufficient level of knowledge overlap to ensure effective communication, interactions between individuals who each possess diverse and different knowledge structures will augment the organization's capacity for making novel linkages and associations beyond what any one individual can achieve. They suggested that the more an organization develops individuals' awareness of others' capabilities and knowledge, the stronger will be the organization's absorptive capacity—the ability to recognize new information, assimilate it, and apply it to commercial ends. Thus, innovation in teams requires not only creative thinking, as may be triggered by minority dissent, but also the absorptive capacity to recognize, assimilate, and apply these creative ideas.
>
> The absorptive capacity of a team will be higher when team members participate in decision making. Participation stimulates the exchange and integration of information, ... reduces resistance to change, and facilitates team members' commitment to team decisions ...

Keeping individuals 'in the loop,' making sure that they are well informed about the contributions which others make to their own, individual success, is one cultural feature, then, of a community of enquiry which overcomes the inhibitive effects of social comparison pressures and which encourages individuals to share information to produce an assembly bonus. This feature shows itself in myriad ways, concretely. For example:

- Enquirers are accustomed to acknowledging both their sources and their sub-authorial collaborators in their presentations and publications;
- Some communities of enquiry, varying across the disciplines in complex ways, are actually organized more or less formally *as* teams, with team meetings, leaders at various hierarchical levels, and so on;

- There are ceremonial or ritual occasions when attention is explicitly drawn to the role of team members, group culture, and role differentiation contributing to common goals—as with anniversary and other celebratory occasions, formal reviews, induction of new members, and so on.[46]

Through all these practices and on all these occasions, we recognize, explicitly, that communities are both the products and the facilitators of our individual as well as our joint endeavors and thus make the issue of transparency of interdependence an explicit topic for group recognition.

6.6 Task conflict and relationship harmony

If a shallow consensus organized around values, paradigms and rules is one crucial cultural element in a well-functioning community of enquiry and transparency is a second, then a third such element is surely individuals' culturally endorsed (and institutionally monitored) attitudes toward diversity and the conflict which it has the potential to induce. There is, again, a crucial essential tension that has to be managed properly for the facilitative possibilities—e.g. of an assembly bonus—to be realized.

It will be useful to begin with a distinction between task conflict and relationship conflict. Mannix *et al.* provide a helpful summary:[47] 'Relationship conflict is an awareness of interpersonal incompatibilities.... Task, or cognitive, conflict is an awareness of differences in viewpoints and opinions pertaining to the group's tasks.' So, for example, A and B might be aware that each has a particular, but different understanding of the interpretation and/or relative significance of accuracy in guiding theory-choice. This would be a case of task conflict: there is conflict, because or to the extent that A and B disagree; but the conflict is task-related (rather than personal) because the disagreement is about how their task should be undertaken. On the other hand, A and B might simply dislike one another. This would be a case of relationship conflict, and it might arise whether or not A and B also disagreed in some specifically task-relevant way.

I suggested that there was a management problem when it comes to these two kinds of potential conflict between members of a community of enquiry. This problem arises because, while task conflict can actually facilitate improved collective performance (much as in Section 5.3), relationship conflict can inhibit the realization of an assembly bonus, broadly conceived. As Mannix *et al.* put it:[48]

> Studies show that relationship conflict is detrimental to individual and group performance, member satisfaction, and the likelihood that the group will work together in the future... When group members have interpersonal problems or feel friction with one another, they may be distracted from the task, work less cooperatively, and produce suboptimal products... In contrast with relationship conflict, moderate levels of task conflict have been shown to be beneficial for many kinds of group performance. Teams performing complex cognitive tasks benefit from differences of opinion about the work being done... Teams improve decision quality as they drop old patterns of interaction and adopt new perspectives; the synthesis that emerges from the conflict is generally superior to their individual perspectives...

When it comes to the efficacy of a community of enquiry, then, much will depend on how they manage the experience of disagreement within the group. If individuals perceive themselves as involved in a situation of relationship conflict, then, again, we have a situation where social-dynamic pressures inhibit collective deliberation. But if they perceive themselves as involved in a (not-too-fraught) situation of task conflict, then, in fact, their performance will actually be enhanced.

Unfortunately, perhaps because of attributional processes (see Section 5.8), members of some communities are likely to read any conflict as relationship conflict and, accordingly, to have their potential assembly bonuses problematized. As Mannix *et al.* put it:[49]

> [R]ecent research in this domain... indicates that perceptions of task and relationship conflict are often highly correlated. One reason for this correlation may be misattribution—for example, mistaking a task-focused comment as a personal attack... Misattribution can occur because people experience conflict as negative and infer malevolent intentions on the part of those with whom they disagree. Misattribution may be more likely in newly formed or highly diverse teams, where a lack of deep knowledge of others may cause members to confuse task conflict with relationship conflict ...

What kind of counterweight can we identify to such an inhibitor of the assembly bonus effect? Mannix *et al.* themselves identify trust, but some more specific remarks of theirs are even more promising, I believe. Writing specifically about what they call 'distributed work,' i.e. work where individual contributors are not normally in regular face-to-face contact, they note that 'two of the particular challenges distributed teams must overcome... are directly antecedent to conflict: lack of a common social identity and increased compositional diversity.'[50] This shows, I think, a way forward, through trust toward an avoidance of misattribution and, hence, toward a productive use of conflict, now understood, within the group, as task conflict.

Let me explain.

What Mannix *et al.* are arguing, in effect, is that, when a common social identity is absent, there is mistrust among members of a collective and, accordingly, even objectively task-related disagreements are likely to be interpreted as marks or effects of interpersonal antipathy. And all this results, as Mannix *et al.* put it, in distraction, lack of cooperation, and lost opportunities to realize assembly bonuses. But what if members of the collective did share a common social identity, e.g. with commitment to the same values and, as Kuhn puts it, a great deal else in common. He says:[51]

> A scientific community consists... of the practitioners of a scientific speciality... [who] have undergone similar educations and professional initiations... As a result, the members of a scientific community see themselves and are seen by others as the men uniquely responsible for the pursuit of a set of shared goals, including the training of their successors. Within such groups communication is relatively full and professional judgment relatively unanimous.

Or, in other words, within such communities, there is a basis for trust and, accordingly, for the kinds of (on Kuhn's account, short-lived) task conflicts that do not degenerate (as they might in different sorts of communities) into relationship conflicts, with all their attendant inhibitions on the pursuit of assembly bonuses. The transparency of potential mutual benefit is also relevant, of course. Insofar as individuals see their own success as depending on the contributions of others, they will have some reason not to infer relationship conflict from task conflict.

It is worth saying a bit more here. What Kuhn and others provide—and the observations generalize, I think, far beyond science *per se*—is an account of a community of enquiry in which opportunities for realizing assembly bonuses have been facilitated, and perhaps even ensured. In particular, and as already noted (Section 6.1.5), there is a shallow consensus about values among members of such a community. Members of the scientific community may differ in how they interpret the values relevant to comparing rival scientific theories and they may differ in how they weight these values when they do not all point in the same direction. But such individuals *do* agree, more or less as a condition of their entry into and continuing membership of such a community, that *these* are the values in terms of which rival theories are to be assessed. (They agree about a great deal more, as well, as we have seen in Section 6.2.) When someone addresses them about the merits, relative to these values, of rival theories, that communication is therefore unlikely to be interpreted in terms of relationship conflict. It is, in the argot of *The Godfather* films, 'not personal; just business'; it is, moreover, precisely what we expect of our 'team mates' in a collaboration for mutual benefit. Accordingly, there is, in such an intervention, no trigger for the reactions—of distraction, non-cooperation, etc.—that are associated with relationship conflict.

Or, in other words, a scientific community, through its initiation rites and practices, is one in which the misattribution of task conflict to interpersonal hostility is minimized, as it can be, of course, in any community where the educational approach which Kuhn describes is an appropriate one—any community, in other words, where a standardizing, rather than an individualizing educational *telos* is apt. (See Section 6.1.4)

6.7 Attitudes toward risk

In any effective community of enquiry, there is, as I've already indicated (Section 2.1), an essential tension between conservation and innovation. On the Kuhnian model I've been elaborating, some individuals will be more and some individuals will be less *risk-averse* when it comes to their enquiries.[52] Indeed, it is necessary, if risk is to be spread, that some individuals, in developing a line of enquiry, will be relatively conservative, continuing to work within an existing framework even in the face of difficulties, and that other individuals will be relatively less conservative, especially in being readier to abandon an existing framework when it become *too* difficult to persist with it.

We could, of course, treat differential risk-aversion as a primitive postulate in our model of communal enquiry: it is, luckily!, simply a fact that there are these differences and that's all we can or need say on this subject. We could, on the other hand, consider what conditions might explain or support the existence of diversity in attitudes toward risk.

And such conditions are, indeed, readily enough identified. John Payne provides a helpful introduction:[53]

> March and Shapira have focused on how attention can shift between a survival value and an aspiration level, also called the *performance target*. They argue that much of the shift in attention is due to changes in a decision maker's resource levels. More specifically, they argue that if a person is just above the performance target in terms of resources, the primary focus of attention will be on avoiding courses of action that might result in falling below the target. Thus, one would see relatively risk-averse behavior. However, as a person's resources increase, the chances that any given uncertain action will result in an outcome that will place the decision maker below the target becomes less. Thus, the person is more willing to take that action (risk). By contrast, if a person has or anticipates a resource level that places him or her below the target value, the person will take greater risks in order to have a chance of reaching the target. The further away from the performance target, the greater the risk taking. When resources fall to a level near the survival point, however, a person will become very risk averse, avoiding any chance of falling below the survival level of resources. Thus, variation in one's resource level can lead to apparent variable risk preferences (attitudes).

Let me explain the relevance of this to communities of enquiry.[54]

A community of enquiry is not constituted all at once.[55] Members join the community at different times. Individuals who have been members of the community for some time will have had more opportunities, than newer entrants, to have acquired 'cultural capital,' as Pierre Bourdieu styled it.[56] They may have acquired standing or recognition in the community, for example, or, more crudely, relatively secure, relatively high-status positions. In other words, there will be a range, across the community, in the level of resources which various individuals hold. Where an individual's resources make him somewhat but not completely secure, he will, on the account of March and Shapira (as reported by Payne), be less inclined to take risks of a kind that would reduce his standing or recognition in the community below some threshold level. Such an individual will, in Kuhn's terminology, be an ideal contributor to the conservative elaboration, normal-scientifically, of the community's paradigm of enquiry.

On the other hand, where an individual's resources are already below, but, perhaps, not too far below, some personal tolerance threshold, she may be considerably more willing to take risks, and thus be readier to be an enthusiast for theoretical or experimental developments that stand in opposition to the dominant framework in her community. Similarly, if an individual's position is already *very* well established, he too may be willing to engage in more risky behavior.[57]

Differential risk-aversion is crucial to risk-spreading, which is, in turn, crucial to effective deployment of a paradigm of enquiry. But differential risk-aversion itself is a structural feature of any community in which accumulated rewards for effective performance are partly correlated with time of entry to the community (and, of course, in which individuals enter the community at different times). It is, if you will, an endogenous, not an exogenous, variable in the system. As Bourdieu put it, 'The structure of the distribution of scientific capital is the source of the transformation of the scientific field through the intermediary of the strategies for conservation or subversion of the structure *which the structure itself produces*.'[58] We do not need to *posit* differential risk-aversion; we can, in fact, explain it and, where necessary or appropriate, plan for it—i.e. design it into new institutional arrangements. (This is, of course, another way in which temporality is important in human affairs. See Section 1.2.)

6.8 The performance loop

How well enquirers perform is partly a matter of their confidence in addressing the problems they are engaged with, and how confident enquirers are partly depends on how successful they have been in solving such problems. Richard Moreland and John Levine provide helpful commentary:[59]

> [A] group's ability to identify problems may also depend on its level of *performance*. Groups that consistently succeed at achieving their goals often become complacent and that complacency can delay the detection of problems. People who belong to such groups rarely expect problems to arise, so they overlook or ignore many symptomatic events...

Translating this into the Kuhnian idiom, we might say that normal science puzzle-solving leads to complacency which, in turn, reinforces a (risk-averse) conservative orientation to the community's dominant paradigm, thus reproducing normal-scientific modes of enquiry. On the other hand, too little success in problem-solving can also inhibit performance, as Moreland and Levine point out:[60]

> [G]roups that are consistently unsuccessful at achieving their goals ... may have difficulty identifying problems as well. People who belong to unsuccessful groups have so many problems already that they are reluctant to discover any new ones.... And they are so busy trying to solve old problems that new ones may develop without being noticed.

This pattern might be characteristic of inter- or pre-paradigmatic science when, on Kuhn's account, there is too little underpinning individuals' enquiries—too little accumulated collective cultural capital (in the form, typically, of a well-established paradigm)—to support effective problem-solving.

Sometimes, however, the problems being addressed are just hard enough for individuals to acquire kudos for solving them and not so hard that they can't be solved. Where this is the case, we have, if you like, a positive feedback loop which sustains the accumulation of cultural capital, and, accordingly, the confidence required to address and to solve problems. How likely this is to happen, and under what circumstances, is a matter which I address in Section 8.4. At this point we can simply say that, evidently, there are some domains of enquiry which do provide opportunities for individuals to contribute to collective enquiry... and that the evidence for their reality is their continued existence as traditions of enquiry—e.g. physics, chemistry, economics, even philosophy: the problems are hard, but not too hard, or, if they are too hard, they are not so hard that even risk-takers are deterred from pursuing them.

6.9 The wave model, revisited

I have mentioned, already (see Section 6.1.3), that the dialectic between diversity and convergence is sometimes played out through a 'wave' or 'cascade' as it passes through a community of enquiry. I'd like to describe, drawing on some materials from my book *Free Public Reason*, how this might work.[61]

First of all, any community of enquiry has a twofold task: to produce knowledge and to make judgments about their own activities (as part of the process of producing the knowledge they produce). Knowledge-production is guided by the values that constitute the standards for achievement and, of course, by the paradigm and heuristics that are current in the community. Secondly, each member of this community takes herself to be trying to interpret the paradigm not just for herself but for the community as a whole. She is what Charles Taylor called a 'strong evaluator.'[62] Her concern is not about 'what should I think?', but, rather, about 'what should we, collectively, and based on our standards of enquiry, think about the matters we are investigating?' Thirdly, each member of the community is committed to evaluating her own (and others') interpretations or articulations of the paradigm in accordance with the values current in that community (and the objects of a shallow consensus). Fourthly, the members of the community of course differ in the ways in which they interpret and balance the sometimes competing demands of the various values, rules, and iconic achievements, so that different individuals, with the same commitments, might nevertheless judge different articulations of the paradigm differently.

Given these familiar points, we can observe that, sometimes anyway, a 'wave' of judgments will progressively sweep through the community of enquiry, as follows.[63]

On account of his particular interpretation of this system of interpretanda, A judges, at time t_1, that one articulation of the paradigm, say T_1, is better than another, perhaps devised by B, say T_2. Accordingly, A works on T_1 (and B doesn't) and it may happen that, by doing so, A so improves T_1 that, at some later time t_2, B now does have reason, despite her differences in dealing with these interpretanda, to prefer T_1 to T_2 (or, anyway, T_1^* to T_2^*, where the * marks the fact that both T_1 and T_2 have been modified in the course of these activities). In this case, A and B now become a team working on T_1^* and, through their joint efforts, they may so improve T_1^* that it now, as T_1+, becomes preferable even for C, who did not previously have, on account of his different understanding of the *interpretanda*, adequate reason to prefer T_1 (and its successors) to T_2. And so on. (Or not, as the case the may.) When this *does* happen, we have a rationally mediated 'cascade' (rather than the 'stampede' that Sunstein sometimes imagines), as follows:

Table 6.3

	A	B	C	D	E
t_1	$T_1>T_2$	$T_2>T_1$	T_1 not preferred	T_1 not preferred	T_1 not preferred
t_2	$T_1>T_2$	$T_1>T_2$	T_1 not preferred	T_1 not preferred	T_1 not preferred
t_3	$T_1>T_2$	$T_1>T_2$	$T_1>T_2$	T_1 not preferred	T_1 not preferred
t_4	$T_1>T_2$	$T_1>T_2$	$T_1>T_2$	$T_1>T_2$	T_1 not preferred
t_5	$T_1>T_2$	$T_1>T_2$	$T_1>T_2$	$T_1>T_2$	$T_1>T_2$

This, I think, represents a common development in communities of enquiry.

6.10 Interim summary

We have, I think, the outlines of a culture of enquiry which might, along with other mechanisms such as multiple accountabilities and novelty premiums, ensure that a community of enquiry is able to function effectively in the face of such social comparison dynamics as might otherwise undermine the achievement of those assembly bonuses which are promised by and implicit in a collective approach to enquiry. On the one hand, we have the shallow values consensus with its potential for risk-spreading variability, the achievement of agreement via dominance, and the residual diversity that maintains the potential for change and improvement even after (temporary, always temporary) agreement has been achieved. On the other hand, we have the trust, to use the vocabulary of Mannix *et al.*, that is built on common values and a common experience of initiation into the community and which facilitates relationships in which the default understanding of conflict is that it is task-related not interpersonal.

We can, I think, now somewhat complicate the parallelogram of forces which I presented a simplified version of at Section 5.9. I will suggest some further complexities in Chapter 8. In the meantime, I would like to discuss something which a great many social epistemologists have mentioned, but which none of them, in my view, has so far adequately analyzed—namely, the division of labor in communities of enquiry.

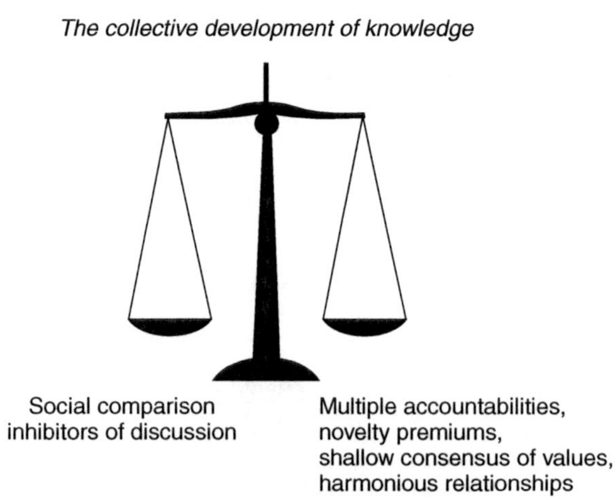

7
The Division of Labor in Communities of Enquiry

In influential work, Philip Kitcher set an agenda and established a method for understanding some ways in which individuals' diversity might be tapped to enhance the production and assessment of knowledge.[1] His account of 'social epistemology' remains a powerful point of departure. He said:[2]

> The general problem of social epistemology, as I conceive it, is to identify the properties of epistemically well-designed social systems, that is, to specify the conditions under which a group of individuals, operating according to various rules for modifying their individual practices, succeed, through their interactions, in generating a progressive sequence of consensus practices.

Kitcher had in effect taken up, much as I have here and elsewhere, some earlier remarks by Thomas Kuhn about the role which diversity might play in the formation of knowledge in communities of enquiry. In particular, Kuhn's so-called 'risk-spreading argument' clearly articulates a specific situation in which 'a group of individuals... [might] succeed, through their interactions, in generating a progressive sequence of consensus practices,' particularly by spreading the risk of adopting a new theory by having only some members of the community do that (while others remain committed to the existing theory). (See Sections 2.1 and 6.1.2.) Kuhn says:[3]

> [I]ndividual variability in the application of shared values may serve functions essential to science. The points at which values must be applied are invariably also those at which risks must be taken. Most anomalies are resolved by normal means... If all members

> of a community responded to each anomaly as a source of crisis..., science would cease. If, on the other hand, no one reacted to anomalies or brand-new theories in high-risk ways, there would be few or no revolutions. In matters like these the resort to shared values rather than to shared rules governing individual choice may be the community's way of distributing risk and assuring the long-term success of its enterprise.

It is this possibility—that individual enquirers might evaluate and engage with problems and innovations in a diversity of ways—that Kitcher explored in a thorough, precise, and convincing manner in his major work in this area.

Kitcher's original paper in this area was called 'The Division of Cognitive Labor.' By the time of his 1993 book, a similar passage was headed 'The Organization of Cognitive Labor.' For cognate material he used the phrase 'Organizing Enquiry' in his 2001 book. This change of wording is significant, I think, because, in fact, Kitcher never really dealt with a proper *division* of labor, in the sense of Adam Smith[4] or Karl Marx.[5] Neither did I in earlier work, though I have made a start, I think, in Section 3.4. What both of us (and others such as Alexander Rueger[6]) were interested in was the way in which various individuals, having formed different evaluative attitudes toward a particular issue, might adopt different stances toward that issue—e.g. might pursue the innovative or instead the conservative approach to a particular problem. What none of us did was consider whether, for example, *a project of enquiry might be sub-divided into different tasks that could then be distributed across different individuals or teams within the community*. (This is, of course, the idea of 'decomposition' which I discussed in Section 3.4.)

But that, of course, is what's required for a genuine *division of labor* in a cognitive or epistemic context. It is also, of course, what we observe in communities of practice concerned with disciplined enquiry—different members of these communities play different roles, perform different functions, undertake different activities and what one person does is meant to be complementary to what others do. In any event, I am going, in the remainder of this chapter, to sketch a framework in which we could talk about the ways in which it might make sense to divide labor in the way, for example, that it is divided in the modern factory or in contemporary design and engineering projects. I will do so in what may seem a rather circuitous way. I will consider, initially, some models from the management sciences, from biology, and from design theory. I will return, in due course, to the issue of enquiry in a more general sense.

What will emerge, really, is a new challenge to collective approaches to the development of knowledge. This challenge, however, arises, not from the difficulties of overcoming individuals' hesitancies in sharing information, but, rather, from the structure of the problems individuals and groups work on. It is, in some cases at least, the ways in which problems are situated in a larger field of enquiry that may make it problematic to approach these problems via a division of labor. Or that, anyway, is what I will argue and then assess. If I am right about this, then there are, as it were, 'objective' as well as 'subjective' impediments to effective enquiry. (What is especially interesting, I think, is that both these kinds of difficulties ensnare our enquiries in path-dependencies.) Again, of course, I draw attention to the difficulties not because they cannot be overcome, but, rather, simply to hammer home the point that where you have an effectively functioning community of enquiry, it does function effectively precisely because or to the extent that such difficulties have indeed been overcome. Much as we might overcome social comparison inhibition by institutionalizing novelty premiums and multiple accountabilities, so we will have to overcome these objective difficulties of effectively dividing labor in some way or another.

7.1 Background

Those contemporary discussions of division of labor which recommend themselves to the philosopher are not, alas, part of the philosophical literature even loosely construed. (The only 'hits' on 'division of labor' in *The Philosopher's Index* that are relevant to the concerns addressed here are to Kitcher's paper,[7] and they are not particularly numerous.) The most interesting discussions, then, are from what we might call, broadly, the management sciences literature.[8] Most of these works share origins in or at least an orientation to the work of Herbert Simon.[9] The reason, in the form of a 'slogan,' is obvious enough: *Division of labor is a response to complexity given bounded rationality*. (See Sections 2.3 and 3.1.) It will be useful, however, to expound this slogan; doing so will enable me to present some apparatus that will prove useful and to associate the analysis presented here with broader issues of more general significance.

I will proceed as follows. First, I will consider complexity and its relations to the kind of evaluative pluralism that is crucial, for example, to Kuhn's axiology. Second, I will consider some of the difficulties of finding better solutions or theories in a 'search space' which is underpinned by complexly pluralized evaluative standards. Finally, I will introduce division of labor as a response to the difficulties of identifying better or best theories or solutions in the face of these kinds of evaluative complexities. I follow on by considering some of the difficulties which a division of labor itself raises, including, especially, the difficulty of designing or identifying an effective division of labor. As it transpires, it is not a trivial exercise to decide how to 'decompose' a complex problem so that delegated teams can work together harmoniously and coherently.

7.1.1 Complexity and evaluative pluralism

As it turns out, the kind of complexity that might ground a division of labor is closely related to the sort of deep evaluative pluralism articulated by Isaiah Berlin and John Gray and already considered in Section 6.1, where I identified it as a crucial feature of the broadly Kuhnian account of disciplined enquiry. The basic ideas are easily summarized as follows.[10]

1. There are, in many domains of evaluation, a *plurality* of different standards against which objects belonging to that domain need to be assessed.
2. These standards will, in many cases, be *interdependent*, in the sense that modifying an object in a way which changes its value against one standard may also change its value against another standard—and not always in the same 'direction'; it will often, perhaps even characteristically, be the case that an object which is made better in one relevant respect is, by that very fact, made worse in another also relevant respect. (Again, this is anticipated by Kuhn.)
3. Because of this, we cannot identify the *overall* best object belonging to this domain simply by asking which object is best against each of the separate standards. There may, in fact, be no *single* object which *is* best on *all* the standards.[11]
4. In many cases, then, superiority against one standard will need to be 'traded off' against inferiority against a second standard in order to (perhaps only partially) determine overall superiority. This will involve what, in a previous book,[12] I called a project of COMMENSURATION, in which we 'price out' evaluative standards against one other, developing a common denominator in terms of which overall judgments of value can be made.

All this, but especially the second point, is resonant with the biological idea of *epistasis*, which is, in turn, crucial to the definition of *complexity* in the technical sense which that term has recently assumed across a range of academic discourses. (See, of course, Section 3.1.6.) Koen Frenken provides a clear exposition:[13]

> In biology, the distinction between the genotype and the phenotype is crucial to the understanding of evolution through natural selection. Mutations occur in the genotype while natural selection operates on the ensemble of traits constituting the phenotype. Traits in turn affect the survival and reproductive success of a phenotype through which its genome (or part of it) is transmitted to its offspring. Epistasis between genes implies that a mutation may not only change the functional contribution of the mutated gene to the entire phenotype, but it generally affects the functional contribution of epistatically related genes to the phenotype as well.... *Due to these dependencies among genes, a mutation in a single gene may have both positive effects on some traits and negative effects on other traits.*

In the presence of epistasis (or interdependence of standards), the ease with which or indeed the very possibility that we can identify an overall best object in a given domain of evaluation will depend, among other things, on the number of relevant standards and the degree to which they exhibit these sorts of interdependencies.[14] As Daniel Levinthal put it:[15]

> The complexity here stems primarily from the often unknown nature and magnitude of interactions between different parts of the system and the consequent system performance implications. The nature of the interaction between two parts may be positive (increasing in one another), negative (decreasing in one another), or unrelated. Furthermore, the nature of the interaction may alternate between positive, negative, and unrelated over different ranges of interaction strength. As a result, *overall system performance can exhibit highly nonlinear and/or nonmonotonic behaviour in response to changes in one or more parts.*

Stuart Kauffman has provided an influential model of these phenomena in which the idea of 'NK fitness landscapes' is deployed and investigated, and of which Lee Altenberg's paper represents a useful generalization.[16] According to these models, the complexity of the evaluative space defined by N standards with K interdependencies among them is linear in N and exponential in K, so that such a domain of evaluation is of magnitude N^K. Epistasis, in other words, generates evaluative complexity in the sense that the evaluator, should she wish to consider the various options, will need to do so against quite a large and complexly structured space of possibilities in which it will not be easy to discover the option, among those she might consider or construct, which is best overall.

Suppose, to make matters concrete, that there are 10 distinct standards in a particular domain of evaluation and that these standards are epistatically linked, e.g. by 7 interdependencies for each of the 10 distinct standards. Suppose too that we want to identify an overall best option in that domain. We might look at a particular option O and consider its value against one of these standards, S_1. Perhaps it rates rather poorly against that standard. We might try to improve it in some way, by transforming it to a successor, O′ so that it, O′, rated better against that standard. But if S_1 is interdependent with seven other standards, then improving the option against S_1 might degrade its performance against one or more of those other seven standards. Note, by the way, that these changes in value can 'cascade' in complex ways. An improvement against S_1 may degrade performance against S_2, but degraded performance against S_2 may be linked, epistatically, to improvement against S_3, and so on. It will, accordingly, be no trivial matter to determine whether the transformation of O into O′ has led, in fact, to an *overall* improvement. Making such a determination will require us to develop or apply a system of commensuration, or, in other words, a system of rates at which the standards trade off against one another in relation to the options for choice. It might require, in other words, that we 'price out' the losses with respect to one standard against the gains with respect to another.

Notice, furthermore and as Levinthal reminds us, that the ways in which standards are interdependent may be quite complex. Perhaps, at the specific locus at which the transformation from O to O′ leads to an S_1-improvement, S_1-improvements lead, in turn, to S_2-degradations. Even so, there may be other loci where transformations leading to S_1-improvements also lead to S_2-improvements. Indeed, at other loci, the ways in which interdependencies cascade may also differ and so on.

Another proviso. We should not, of course, take too seriously the idea that an evaluative space is 'given' and simply needs, for us to locate an optimum, to be traversed—however that might be arranged. (I return to this matter periodically and, in a reflective mode, in Section 7.4.) The options occupying various loci in such a space may, in fact, have to be constructed, much as we imagined the construction of O′ from O. And it may not be evident, indeed, what the trade-offs among standards are until the various options have been constructed. Our preferences over options (and the trade-offs which underlie them) may themselves be constructed rather than taken as given (or even discovered). This point will prove to be significant and reflects, I believe, some key ideas of decision cycles or constructed preferences theory.[17] It is certainly implicit in my idea that, in comparing options, we may sometimes need to establish first a mechanism for doing so . . . and then, to effect the comparison, apply that mechanism.

By the way, in distinguishing, implicitly, between the transformation of an object and consequential changes to its overall worth on a high-dimension evaluative space, I am, via Frenken,[18] following Saviotti and Metcalfe,[19] in distinguishing between 'technical' and 'service' characteristics of the objects under assessment. For example, when we transform an option O into a successor O′, we do so by changing something about that option. If it is a theory, then perhaps we add a parameter; if a physical object we are designing, then perhaps we adjust one of its characteristics—e.g. size, color, etc. And, by making these 'technical' changes, we affect the value (or its 'service characteristics') in various ways. Perhaps adding the parameter to the theory enables us to make more accurate predictions within a certain range. If predictive accuracy is an evaluative criterion for theories, then this change in technical characteristics induces a change in the service characteristics or, in our terminology, in the value, along a significant dimension, of the option in question. (Of course, adding a parameter may make the theory less simple and, if simplicity is also a standard for the evaluation of theories, this interdependency contributes to the difficulty of making an *overall* evaluation of the options.)

7.1.2 The geography of complexity

When we have complexity of the kind that communities of enquiry typically do face, it will, it transpires, be problematic to organize enquiry so that the best option (e.g. theory, account of the results of some experiment, etc.) might be effectively and efficiently identified by the members of the community. (This point is independent of, but is certainly strengthened by, the possibility that trade-offs and preferences (and options to be evaluated in terms of them) are constructed rather than given.) This is a way, in effect, of characterizing the boundedness of our rational engagement with problems or enquiries subject to this kind of complexity. To bring home the difficulties, for finite beings, in relation to complex enquiries, we might consider the 'geography' of the 'spaces' in which such enquiries are conducted. Marengo *et al.*, adapting Kaufmann's NK terminology and apparatus, provide a clear statement:[20]

> The value assumed by K is a key point with respect to the shape of a fitness landscape and, consequently, to its exploration. Indeed, for K = 0 the contribution of each element to overall fitness will be independent of every other element... The highly correlated structure of the fitness landscape [for K = 0] can be effectively exploited by local and incremental search processes. Conversely, *as K increases, the landscape will be increasingly rugged and points that are close in the landscape will no longer have similar fitness values.* What is most relevant to our point is the fact that the more rugged the landscape, the less (locally) informative is its exploration...

The key idea, then, is that, when a community engages with an enquiry which is subject to evaluative complexity, it will encounter a 'problem space' which is 'rugged' in a sense which can defeat effective and efficient systematic enquiries across that space. The difficulty is, to be specific, that there will be numerous 'peaks' of overall value more or less randomly scattered over the (of course multi-dimensional) space and, especially as interdependencies increase, little to distinguish among them. (It is important to note that our analysis presupposes that we *have* COMMENSURATED the standards. Difficulties navigating a complex landscape are, of course, as I have pointed out and will reiterate, made more difficult when the landscape itself is constructed 'as we go,' as, of course, it almost always is. The difficulties we are examining here, however, do not depend on this point. *Even a fully constructed evaluative landscape will be difficult to search for the best option when that landscape is rugged.*)

Frenken provides a useful summary of some of the key points about the geography of complex landscapes of enquiry, such as those that typical communities of enquiry might encounter:[21]

1. The number of local optima increases exponentially with complexity *K*....
2. The... higher the complexity of a system, the more randomly spread the local optima are in design space.
3. ... When K-values get larger, they pose increasingly more incompatible design constraints on the design, resulting in poorer performance

of local optima (with expected fitness equal to 0.5 for complexity K approaching infinity).

4. ...
5. The variance of fitness levels of local optima falls as complexity K increases.

A complex space of enquiry, then, has a large number of rather mediocre 'local optima,' scattered 'randomly' across the space. The difficulties of navigating such a space can now be exhibited by considering some enquiry strategies. Any such strategy will have to identify (a) where the enquiry begins, (b) how the enquiry proceeds, and (c) what the 'halting' routine is for this enquiry. As is well known, two rather obvious enquiry strategies each fail to satisfy important meta-criteria when applied to problem spaces that exhibit the kinds of evaluative complexities which typical communities of enquiry must in fact deal with.

7.1.3 Enquiry and complexity

One way of characterizing the difficulties raised by complex problem spaces is to contrast the relative advantages, as enquiry strategies, of *exhaustive* and *myopic* (and, later, *parallel*) enquiry across such spaces. (For exhaustive and myopic enquiry, I adopt the terminology of Frenken.[22])

For exhaustive enquiry, the strategy is: (a^E) begin the enquiry at any randomly chosen point in the problem space; (b^E) make any change or combination of technical changes to the object whose value is being considered and record the effect of that change on the overall (i.e. service) value of the object; (c^E) halt the enquiry when all combinations of changes to the object have been surveyed and identify, as the optimum combination of technical characteristics, that combination whose associated overall value is a maximum in relation to all others.

For myopic enquiry, on the other hand, the strategy is: (a^M) begin the enquiry at any randomly chosen point in the problem space; (b^M) make any single-trait technical change to the object whose value is being considered and consider whether this change raises, lowers, or fails to affect the overall (service) value, on this space, of this object; if the change increases the overall value of the object, proceed, iterating step (b^M), from this point, otherwise from the original starting-point; (c^M) halt the enquiry when there is no change to the technical characteristics of the object which does not lower the overall (service) value of the object.

Exhaustive enquiry should identify the global optimum even in a complex evaluative space. (Notice that we are presuming the existence and identity of a global optimum and, hence, of a completed project of COMMENSURATION.) It will require, however, that our enquiries range across the entire space, which, when N and K are large, will itself be extremely large—for $N=10$ and $K=9$, the dimensionality of the enquiry space is 10^9. This means that exhaustive enquiry will be extremely resource intensive; it will require considerable time and creativity on the part of the community of enquiry... *too much* time and creativity, in many cases, for boundedly rational agents. (All the difficulties of enquiry across a preexisting NK evaluative landscape are only compounded, of course, when we have to construct it as we go—develop variants on existing options and COMMENSURATE standards of evaluation which have not already been 'priced out' with respect to one another. This latter project is, of course, much more like what really happens in concrete and specific projects of enquiry. Certainly, the 'space' in which enquirers typically move is not a preexisting one.)

On the other hand, myopic enquiry is more likely to 'halt' in some reasonable time and hence be suitable for boundedly rational agents, but only because it is more likely to do so at the nearest local optimum from its starting-point, with the likelihood, tending asymptotically to certainty as K increases, that this local optimum is *not* the global optimum for the system as a whole. As Frenken says:[23]

> [M]ost adaptive walks will end up in local optima with sub-optimal fitness. Myopic agents simply climb up the nearest hill without knowledge of other peaks with higher fitness. Only agents whose initial conditions are, by chance, with the *basin of attraction* of a global optimum..., may end up in the global optimum.... The... higher the complexity of a system and the larger the size of a system, the more likely agents are to become trapped in local optima.

We have a (meta-level) trade-off, then, between these enquiry strategies for complex evaluative spaces.[24] As Frenken puts it:[25]

> Although exhaustive search will always find the global optimum, yet to be fully certain to have the global optimum a designer requires 2^N trials (the size of the entire design space). Exhaustive search is the most inefficient search strategy in terms of costs of search. By contrast, applying a search distance of 1 [as with the myopic strategy] runs the highest risk of ending up in a sub-optimum, yet is the most

cost-efficient strategy. A trade-off exists between the effectiveness of search regarding the result in terms of the fitness of the solution and the efficiency of search in terms of the time required to reach this solution.

7.2 Division of labor

Marengo and Dosi provide a succinct and cogent formulation of an obvious possibility, given the analysis to this point.[26]

> [A]s pointed out by Simon (1969), problem-solving by boundedly rational agents must necessarily proceed by decomposing any large, complex and intractable problem into smaller sub-problems which can be solved independently, i.e. by promoting what could be called *the division of problem-solving labor.*

In other words, the mistake at the heart of both exhaustive and myopic enquiry strategies has been to treat the problem space as an indivisible whole. If the problem space is large and complexly structured and we simply attack it 'head-on,' then, of course, myopic enquiry will be ineffective (fail reliably to identify the global optimum) and exhaustive enquiry will be inefficient (fail to identify the global optimum in a timely fashion). The lesson, clearly, as Simon already saw, is not to attack such a problem 'head on,' but, rather, to divide it into parts (see Section 3.4), allocate the parts to teams, allow the teams to solve the resulting sub-problems, and then assemble the solutions discovered by these teams to provide an overall solution to the problem in question. (And this is, of course, the relevance, potentially, of these materials, to the issue left unaddressed by Kitcher, and by me in earlier work—the division of cognitive labor, e.g. in epistemic communities.)

There are, of course, problems which would readily yield to such an approach. They involve problem spaces which are 'decomposable' in the specific sense that they 'can be partitioned into subsystems such that all epistatic relations [i.e. evaluative interdependencies] are located within subsystems and not between subsystems.'[27]

Where a system is decomposable, a *parallel* enquiry strategy might be deployed as follows: (a^P) for each independent subsystem, begin the enquiry at any randomly chosen point in that portion of the problem space; (b^P) make any single-trait 'technical' change to the object whose value is being considered and consider whether this change raises, lowers, or fails to affect the overall value, on this space, of this object; if the change increases the overall value of the object, proceed, iterating step (b^P) from this point, otherwise from the original starting-point; (c^P) halt the enquiry when there is no change to the technical characteristics of the object which does not lower its overall (service) value; (d^P) assemble the objects identified as embodying local optima within each subsystem to constitute a globally optimum composite object for the system as a whole. Each constituent object has been locally optimized against those standards which are relevant to its evaluation (and which are, by hypothesis, not relevant to the evaluation of any other constituent object—this is what decomposability *means*). When combined, then, these objects constitute a composite which exhibits precisely the overall best balance of service characteristics.[28]

While such a parallel enquiry strategy represents a considerable improvement in terms of efficiency and effectiveness over the exhaustive and myopic strategies, it too is compromised—in its case, of course, by the relative rarity of genuinely decomposable systems of epistatically related evaluative standards. Here the graphical representations which figure so effectively in the analysis of Carliss Baldwin and Kim Clark[29] will be helpful. Their example is of interdependency in design factors for a personal computer. And what their diagram shows is that, while it is possible to partition interdependencies in rough correlation with functionally differentiated sub-systems of the computer—e.g. drive system, main board, etc., there are, as they put it, 'out-of-block' interdependencies, where, for instance, some feature of the drive system is influenced by and influences the functionality of the main board, or of the screen. So, for example, one variable, A might be data-transfer rate. If B were bundle size for data transfer, then an **x** in Baldwin and Clark's cell AB indicates that transfer rates and bundle size are interdependent... and the **x** s in other cells in column A and row B indicate the presence of still other relationships of interdependency.

What such a diagram illustrates is that, while interdependencies might 'cluster' around certain component subsystems within a larger integrated system, there will nevertheless often be, as there are in this case, what we might call 'residual' interdependencies which will (or ought to) affect the design of these component subsystems. As Baldwin and Clark put it:[30]

> In fact, there is no way [in their example] to divide the overall design into independent [strictly 'decomposable'] subdesigns. If by some chance the drive system, say, were designed independently of the main board, the matrix indicates that there are twenty-seven ways

> that the resulting designs might be in conflict (27 out-of-block **x** s). Any one of these dependencies, if unaddressed, could lead to the catastrophic failure of the whole design ...

In other words, the parallel enquiry strategy outlined above, if run on the four subsystems identified in the diagram, could not be guaranteed, as it could be in a *fully* decomposable system, to deliver the global optimum. It cannot be guaranteed, in other words, that a composite object formed from four subsystems which were optimized locally would itself be either technically feasible or globally optimal in its service characteristics. As Marengo and Dosi put it:[31]

> [T]he extent of the division of problem-solving labor is limited by the existence of interdependencies. If sub-problem decomposition separates interdependent elements, then solving each sub-problem independently does not allow overall optimization.

Optimization within one of the component subsystems, treated independently of the others, might fix one or more parameters in ways that, because of the inter-componential interdependencies, were not compatible with optimization either of other components or of the system as a whole. Once again, Marengo and Dosi provide an insightful formulation:[32]

> [I]nterdependencies undermine the effectiveness of the selection process as a device for adaptive optimization and introduce forms of path-dependency with lock-in into sub-optimal states which does not originate from the frictions and costs connected to the working of the selection mechanism, but from the internal structure of the entities undergoing selection.

This will, perhaps, bear some explication. Suppose that, in considering how to optimize the value of the Drive System (DS), we were to make a particular design decision, about, say, how to optimize with respect to the parameter B. Such a decision will of course impact on, and we would not reach it without considering its impact, in relation to overall value, on the decisions available to us on other specifically DS-relevant design parameters. Suppose, then, that we have made precisely that decision affecting the value of B that does indeed optimize, overall, the performance of the Drive System considered as an 'independent' subsystem. (Notice, again, that I am presupposing that these service characteristics have already been COMMENSURATED, and that the relation between technical and service characteristics is well understood.) Notwithstanding such '*locally* global optimization,' we have *ipso facto* fixed or at least constrained the variation in relation to other design decisions that affect other parts of the system as a whole (and, accordingly, the overall value of the system as a whole), precisely because, of course, decisions affecting B are interdependent with other decisions in other component subsystems. For example, making a decision about technical characteristics that fixes B or at least constrains decisions about the parameter H, which is relevant to the performance of the Main Board (MB).

Another way of putting this brings out the difficulty for this sort of division of labor. If the Drive System team optimizes locally, it will fix the value of B in a certain way. If the MB team optimizes locally, it will fix the value of H in a certain way. Since these parameters are in fact interdependent, it cannot be guaranteed, and in some cases will not happen, that the decision which fixes the value of B will be compatible with that which fixes the value of H. Sometimes this will mean that the composite design is not actually technically feasible—the decision we make affecting B means that we cannot physically assemble the Drive System and the Main Board into a working unit given the separate decision we have made affecting H—there is technical incompatibility, we might say. Sometimes, however, even if there is no such brute incompatibility, there may well be sub-optimality. Because we have fixed B and H in the ways we have, we cannot explore the entire evaluative space and, hence, cannot be certain to have encountered exactly that combination of technical features which does optimize the service characteristics of the composite object.

Indeed, this second possibility is what Marengo and Dosi mean when they refer to 'path-dependency.' (See also Section 4.6.) A choice affecting B affects other choices and may result in what they call 'lock-in'—we are locked into choices, for example, about H (and hence about the performance of *another* subsystem) by whatever choices we might already have made affecting a parameter relevant to the performance of the DS. Whatever solution we arrive at to the problem of overall optimization is therefore dependent on the specifics of the path we have taken to arrive at that decision. If we decide on matters affecting B first, then we lock in a decision about H and, accordingly, about what sub-space of the total problem space we can explore in search of an overall optimum. If, on the other hand, we decide on matters affecting H first, then we may well restrict the range of technical choices affecting B that we can make. In this case, we may follow quite a different path in search of some global optimum. In either case, where we arrive is dependent on the order in which we make various decisions about where and how to seek value. (Notice, by the way, that this is unavoidable. It is not an artifact of bounded rationality, but, rather, of the structure of the objective problem situation.)

Baldwin and Clark provide a clear statement of one implication of path-dependency in relation to global optimization.[33]

> Unintended, possibly dire consequences arise when innate interdependencies in the design are not addressed in the task structure of the design process.... In cases like this designers must cycle and iterate through the tasks, searching for a combination of parameters that 'works.' Such cycles are extremely common in design processes. Christopher Alexander describes them in his classic work on design theory, *Notes of the Synthesis of Form*:
>
> *It is a common experience that attempts to solve just one piece of a problem first, then others, and so on, lead to endless involutions. You no sooner solve one aspect of a thing, than another point is out of point. And when you correct that one, something else goes wrong. You go round and round in circles, unable to produce a form that is thoroughly right.*

If parallel enquiry results in 'lock-in,' then cycling is one response. Make a design decision which fixes B and then see what happens to H. If what happens to H is too 'dire,' then reconsider how B has been fixed. If some combination of technical characteristics affecting both B and H seems acceptable, then consider the impact of 'locking in' this combination on other epistatically related variables. If impact is too 'dire' in relation, say, to U (which is two-steps interdependent with B), then reconsider the decision about B/H. And so on. Again, there are difficulties from the point of view of effective and efficient *division* of labor. As Raghu *et al.* point out, in the presence of dense interdependencies, the genuine independence of teams working on subproblems is highly compromised. They say:[34]

> In organizational and social systems—perhaps in mechanical ones as well—it is possible to think of interdependency and interaction among the parts as a matter of information transmission or *communication*. Consider, with Eric von Hippel (1990), the problem of organizing product innovation. Here, the issue is how

> to decompose the organization of a research and development process by partitioning tasks among development teams. As von Hippel pointed out, in order to solve this decomposition problem, one has to focus on the interdependencies among the various tasks the project comprises. If the project is organized in a non-decomposable manner, then interdependency will be high, meaning that each development team will need constantly to receive and use information about what all the other development teams are doing.

Obviously, when the system of interdependencies is 'complex,' such cycling in search of 'a combination of parameters that "works" ' is, potentially, *exhaustingly* iterative. This is not, of course, a happy conclusion to reach. Indeed, we know, experimentally, that this sort of cycling can be very debilitating, psychically and otherwise, for workers engaged with these sorts of systems. As Terry Connolly and Ken Koput put it:[35]

> These [experimental] demonstrations present subjects with environments in which hidden variables are strongly coupled, important patterns build slowly, responses are delayed, and feedback loops are perverse—probably realistic representations of real-world complexity, but cognitively overwhelming. Dorner, for example, reports such dysfunctional behaviors as excessive focus on a single problem and/or sequential problem hopping; failure to cope with slow systems response or long-term trends ('last-period focus'); and response rigidity in the face of deteriorating system performance.... Highly dynamic, complexly looped environments are certainly difficult to understand and operate.

(This may sound familiar to close readers of Kuhn, especially on the pre- or inter-paradigmatic periods of scientific activity, when, for instance, 'different men confronting the same range of phenomena, describe and interpret them in different ways.'[36])

In the face of these difficulties, we seem to be led back, by cycling, in the direction of exhaustive enquiry strategies—already rejected, for complex systems, as too resource intensive relative to our finite powers and resources, and certainly, in view of these demands, no less likely than cycling to induce adverse reactions.

Summarizing, we might say, then, that, for highly complex problem spaces, myopic enquiry is not reliably effective; exhaustive enquiry is likely to be inefficient; and parallel enquiry is likely to be inconclusive, in the sense that there is no natural 'equilibrium' into which it can rightly settle. Graphically we have, if you will, a *meta*-cycle.

7.3 Modularity and population approaches

All this is, if you will, stage-setting in relation to the idea, very influential in design and management circles, of modularity.[37] Of course, in introducing this idea, I do not want it to be thought that I am offering it, or indeed that any of the proponents of this idea are offering it, as a definitive solution to the problem which we have sketched, and which is summarized in Figure 7.1 below. Modularity, if you will, is a way of *addressing* the trade-offs between effectiveness, efficiency, and conclusiveness in divisions of labor; it is not a way of *obviating* the need to make exactly these trade-offs.

I will begin by presenting what is, in the management science and design science literatures, a standard account of modularity. The idea of an *interface* is, I think, the key feature of this standard account. Later,

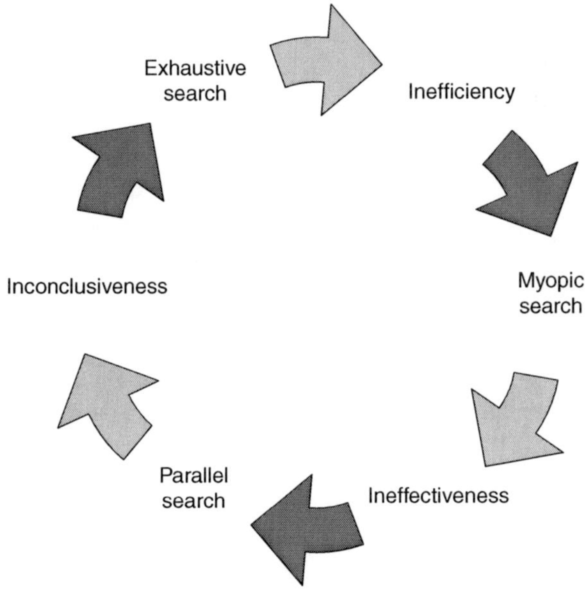

Figure 7.1 A cycle for search-based enquiry

I will consider whether a population approach might play a role in our understanding of complexity and the problems it poses for effective identification of good divisions of labor.

7.3.1 Modularity

We can hardly do better than begin with Ron Sanchez and Joseph Mahoney's influential work. According to them,[38] '*Modularity* is a special form of design which intentionally creates a high degree of independence or "loose coupling" between component designs by standardizing component interface specifications.'[39]

This will bear some explication. As we have seen, the attempt to allocate incompletely decomposed subsystems to separate teams is threatened by the possibility (rising to a certainty when interdependencies are dense) that there will be cross-component as well as intra-component interdependencies and, accordingly, that there will either be mismatches between the work of teams or the need for extensive, ad hoc and iterative communication between teams (thus undercutting the purported advantages of dividing the labor).[40]

One solution to this problem is to modularize, where this encompasses several tasks.

First of all, we try to so divide tasks as to minimize the impact of any remaining cross-component interdependencies. This might involve minimizing the number of cross-component interdependencies, but it might also involve considering the relative strengths of the various interdependencies which we have identified. For example, if one evaluative parameter has a very marked effect on another, then, other things being equal, this would be a reason for including both within a single component.[41]

Another factor we might consider is, of course, the degree to which interdependencies cascade, as I put it earlier. For example, if parameter B influences parameter H which, in turn, influences parameter U, then, depending on how rapidly the influence decays, we might want, *ceteris paribus*, to include B, H, and U within a single subsystem.

Secondly, and crucially, for those cross-component interdependencies which remain after partitioning, we create an *interface* which consists, in effect, of an assignment of parameters in such a way as to ensure consistency when the components are brought together. So, for example, I considered, earlier, how independent design of the Drive System and Main Board might come unstuck given the interdependency of those design parameters affecting the values B and H. If we are modularizing the larger system, we will, then, simply have to make design decisions—which, by implication, *fix* values for B and H—so that the resulting subsystems will prove to be compatible, whatever other decisions are made affecting parameters within these subsystems. Such an allocation is what Baldwin and Clark call a design rule.[42] The subsequent work, on DS and MB subsystems, by the teams allocated those tasks, is subject to this rule and, accordingly, can be expected, other things being equal, to produce, at least in this respect, subsystem designs or solutions which *are* compatible with one another, at least in the sense that they are technically feasible.[43]

A design rule therefore implicitly defines what Brian Loasby would recognize as '[a]n organisation chart... [which] specifies what interdependencies should not be taken into account.'[44]

The advantages of modularization are obvious enough. The problem space *within* any given module is orders of magnitude less complex than the space for the system as a whole. It is a sub-set of the total problem space and, where there would be cross-component interdependencies, these have been eliminated by the imposition of design rules which fix the values of interdependent parameters. Accordingly, an enquiry strategy which halts within a reasonable period of time need not always be myopic and, where it isn't, can discover, with reasonable effort, an optimum for each subsystem. Even myopic enquiries across such a sub-space are more likely to discover an optimum, given the lower evaluative dimensionality of the space. Furthermore, because of the interface created by the design rules, within-module enquiries should yield subsystem solutions that are compatible with one another and whose fit with one another does not need to be negotiated between teams on an ad hoc (and time consuming) basis. Finally, cycling will occur only in the relatively small sub-spaces defined by the various modules, giving a higher degree of conclusiveness.

On the other hand, there are down-side risks to modularization. (Of course, they are simply an internalization, if you will, of the risks already identified when the problem space is complex and undivided.) As Baldwin and Clark themselves point out:[45]

> On the minus side, designers will lose the ability to explore some parts of the space of designs—in effect, the architects will restrict the search, declaring some parts of the design space to be out of bounds. If those areas have already been explored and are known to have no 'high peaks of value,' then the cost of limiting the designers' search in this way will be low. But if the areas have not been explored

> and do contain 'high peaks,' then those superior designs will not be discovered, and the cost of imposing the design rule will be high.

We limit cycling and incompatibilities by taking design decisions which fix certain evaluative parameters that would otherwise exhibit interdependencies. (There is no need to fix them if they don't exhibit interdependencies!) But fixing these parameters means that we cannot explore their relations with one another and, hence, cannot be sure that we have tuned them to achieve an overall best outcome for the system as a whole. It might well be, for example, that the values of B and H that we have imposed, in devising a design rule that enables us to isolate the work of the Drive System team from that of the Main Board team, are not that combination of B and H values that in fact would lead to an overall best design for the computer as a whole. But, of course, we can only tune these values to such a global optimum by engaging in an exhaustive exploration of the entire problem space, which, as we have already agreed, is typically inefficient and, indeed, impossible in real time. (Notice that these downside risks are absent in the case of genuinely decomposable systems.)

All this means, of course, that *modularization imposes path-dependency*. The optimum that is discovered by exploring a modularized problem space is the optimum that you can reach, in other words, along paths in part defined by and hence relative to the design rules which create the interfaces between subsystems. If we had made different choices about the design rules, different optima would have become accessible to search. And since we haven't, by hypothesis, exhaustively enquired over the problem space as a whole, we cannot be sure, antecedently, that we have chosen exactly those values for the parameters covered by our design rules that would lead us along or at least make accessible to us the pathways leading to a global optimum for the system as a whole.[46]

Actually, the difficulties are greater still. Although I have occasionally qualified my account by pointing to the constructed rather than given nature of the problem space (and, of course, of the outcomes of enquiry that we assess), I have nevertheless mostly followed the commonplace diction of management science and design theory in describing the issues before us. It is time to bring these provisos to the fore. In this regard, the work of Daniel Levinthal is especially helpful. As he points out:[47]

> Not only are systems not fully decomposable, but also *apt partitions that capture the nearly decomposable structure of complex systems are not*

> *self-evident*. Designers are searching on the matrix of design parameters at four levels: (1) the 'appropriate' number of modules; (2) the 'appropriate' mapping of design elements to the modules; (3) the 'appropriate' interactions among the design elements within each module; and (4) the 'appropriate' interfaces or interactions between modules.

In other words, in the absence of an exhaustive enquiry over a space that we would, in many cases, be constructing as we went, we cannot tell how to partition the problem space as a whole in the best way. The clustering of interdependencies that enabled us, in the diagram borrowed from Baldwin and Clark, to partition the system as a whole into four relatively independent modules *was not given* but itself had to be discovered, initially in unfamiliar systems, by trial and error, by a process of COMMENSURATION or, indeed, by a process preliminary to COMMENSURATION properly speaking. The loci of interdependency; the strength of interdependencies—these are matters that have to be discovered, often by the constructive exploration of a problem space of high dimensionality. Since, as we have established, such an enquiry will be incomplete and haphazard in complex problem landscapes, any modularization is apt to be based, in varying degrees, on guesswork rather than established facts about these interdependencies and, hence, about how the total system should be modularized. (After all, if we knew the system well enough to know how to modularize, we might well have discovered, already, the global optimum and thus have no need to modularize and enquire!) And, as Baldwin and Clark point out,[48] '[i]mposing a design rule when one is ignorant of the true underlying interdependencies can lead to design failure.' Absent exhaustive (constructive) enquiry, we may even fail to see that an attempted modularization did not manage to control for cross-module technical (let alone service, i.e. evaluative) interdependencies and, accordingly, our teams of workers may produce subsystem solutions which are not compatible with one another even though they conform to the design rules we have imposed.

In modeling precisely this risk, Levinthal has discovered, ironically, that highly modularized systems are more rather than less likely to exhibit such incompatibilities. As he put it:[49]

> [C]onsistent with prior work (Marengo *et al.*, 2000), the effectiveness of innovation is lower, or even disruptive, the greater the deviation of the design structure from the true underlying structure. Second,

and more surprising, we find that, in the long run, erring on the side of greater integration poses lower performance penalties than erring on the side of greater modularity.

The first of these points is obvious enough. The less accurate our assessment of the interdependencies which actually exist (what they are and how strong they are), the more likely the failure of any modularization of this system that we construct on the basis of this assessment. The second point is even more intriguing. In the absence of the kind of knowledge produced only by an exhaustive enquiry that we will not (and perhaps could not) have carried out, a good heuristic principle for modularization is 'err on the side of less rather than more modularization,' or more concretely 'have fewer rather than more subsystems (and hence independently functioning teams of workers).' Of course, with fewer modules, the evaluative dimensionality of the subsystems remains relatively high and, accordingly, myopic enquiry strategies may not discover global optima even within these subsystems (and exhaustive enquiries within these larger subsystems will continue to be costly).

One is tempted to say that we are going round in circles.

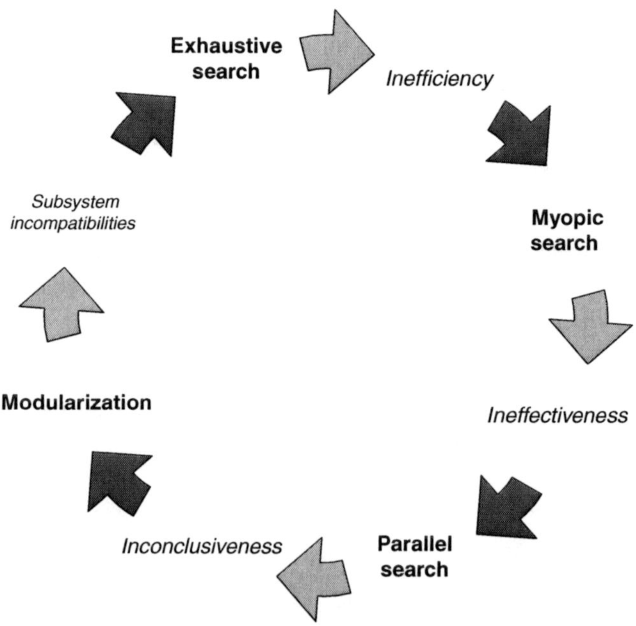

Figure 7.2 A cycle for search-based enquiry

7.3.2 A population approach

Here is where the idea of *diversity* which Kitcher focused on becomes relevant again, though not in precisely the way that Kitcher himself had in mind. When we need, to cope with complexity, to partition the problem space... and when, inevitably, we do not know how to divide the space and allocate the labor, what we should do if we are able (and should try to set the conditions for if we currently are not able) is, as Marengo *et al.* put it, set up 'populations of agents... that compete on the basis of search strategies based on [distinctive] conjectural decompositions.'[50] (See Section 8.2, where I describe a federalist model for enquiry that somewhat resembles, though it is much less formalized than, the model of Marengo *et al.*) As Brian Loasby put it, 'there is no unique way of framing any problem, nor any unique way of operating within any imperfectly specified framework.'[51] What we should do, in other words, is create or nurture a social situation in which different individuals and teams construct different design rules and interfaces and then, in effect, compete with one another to create objects of value, with, perhaps, the team producing the best overall design being recognized for this achievement and then emulated by others who had chosen inferior bases for design. (See, however, Section 4.7, for a cautionary note about emulation.) This will mean, in effect, that we will have a collection of differently implemented enquiry strategies over the same underlying objective problem space (if, again, it makes sense to posit such a space). Each distinctive strategy will divide the space differently and perhaps even explore it differently. (It may even define its fundamental parameters differently.) That strategy which explores the right part of the problem space in the right sort of way will, presumptively, prevail over others.

Such a population of distinctive enquirers will be less hostage to potentially mistaken partitionings of the problem space. And to the degree that they also differ in the ways in which they construct interfaces between (and design rules governing) modules within the larger system, they will be less hostage too to lock-in and path-dependencies. Or, at least, they will, *collectively*, be less hostage to lock-in. Individually, of course, each will be, at least pro tem, locked into her own favored enquiry strategy and to the paths it puts her on.

As Kuhn says,[52] and as both Kitcher and I have tried to show, 'individual variability in the application of shared values may serve functions essential to' rational enquiry. In particular, individual variability in (constructively) partitioning and exploring the multi-dimensional problem spaces by which rational enquiry is typically characterized offer a better response to boundedness and complexity than does a simple division of that problem space. *We need both division and diversity to pursue enquiry in complex situations*. And sometimes, of course, this combination of diversity and division of labor will enable us to identify a global optimum. One of the teams will adopt a system of modularization and enquiry that enables them to identify an option (object, problem-solution, etc.) that is best overall and can be recognized as such.

Notice, furthermore, how this approach differs from the interface/design rules approach we considered earlier. That approach works by defining interfaces in terms which are fixed for all—the various teams can work separately and then bring their products together to form a coherent composite precisely because the design rules specify precise and fixed values for key (interdependent) parameters. The more Kuhnian (or is it Feyerabendian?) approach of Marengo *et al.* works by proliferating competing design rules in which values are fixed locally but vary from one design team to another, thus reflecting, for example, the different interpretations which each of them brings to the paradigm, to the systems of values, and so on. (See Section 6.2.)

None of this is to deny, of course, the persistence, indeed the ineliminability, even when we deploy both diversity and division, of many of the issues which we have canvassed.

One important proviso arises from the simulation studies of Marengo *et al.* themselves, who point out that it cannot be guaranteed, of course, that the objectively correct enquiry strategy will indeed infallibly identify the objectively optimum design in a population of competing design teams. As they put it:[53]

> [I]t is remarkable that correct decomposition strategies [even when represented in the population of competing design teams] might not always prevail when nested in competitive environments characterized by some form of selection. In fact, while they are able always to locate the global optimum with certainty, the time required might be so long that they are actually eliminated by the selection mechanism.

This is, as Marengo *et al.* rightly state, a remarkable, and a chastening, result. In the face of complexity—and that, surely, is our condition in most domains of enquiry—and given our boundedness, we cannot be sure, even when the correct enquiry strategy is represented in a population of enquirers, that this strategy will indeed prevail in competition with other less meritorious strategies.

How could this happen? Here is one way; it is not unrealistic in relation to typical situations in which groups of enquirers compete with one another and attract adherents in accordance with their perceived success.

During any given period of enquiry and at any given region in the problem space, it may happen that a particular enquiry strategy is superior, at that period and in that region, to others. Sometimes this superior strategy will be one which is, in fact, potentially superior both in the longer term and in a wider region. Sometimes it will not, however, have these characteristics. If the competition for survival into the next period is responsive only to local conditions... if the competition for propagation to wider regions is responsive only to local conditions ... then, if the locally superior strategy is not the globally superior strategy, it may still prevail, while the globally superior strategy is not propagated to the next time period or the wider region.[54]

In short, while adding diversity to division enhances our prospects for finding a good solution to a multi-dimensional problem, it does not and cannot guarantee that our solution actually optimizes the trade-offs among the relevant evaluative standards. Here again there is a form of path dependency, I believe. Here the path is traced by the history of the competition between teams (rather than, as previously, by the history of each team's exploration of the space of enquiry). After all, approaches to partitioning and enquiry which would lead to the best design if persisted with may not be competitive, in relation to other, objectively inferior approaches, at every stage of their elaboration. They may therefore be eliminated if unlucky enough to encounter other approaches which are, even if only temporarily, superior to them in whatever ways are relevant to the selection process.

What the simulations of Marengo *et al.* seem to show, in effect, is that we are stuck with history, or, as Dosi puts it, 'the explanation of why something [such as a problem-solution] exists rests on how it became what it is.'[55] There is, in effect, no equilibrium explanation for the solutions that prevail in many domains—no explanation that says, in effect, that, *wherever* you started from, you were in the basin of attraction of the objectively best solution and you were, *unavoidably*, sucked upward toward it. Sometimes, indeed, we will, even in a population of diverse teams of enquirers, be sucked upward toward solutions that are locally or temporarily meritorious without being globally optimal. As Marengo *et al.* put it:[56] 'Suboptimality and path-dependence are... ubiquitous features of ... problem-solving.' C'est la vie!

I mentioned, in Section 5.5, that, on the kinds of pluralist models I have been working with, we could generate the diversity needed to facilitate risk-spreading without invoking sordid motives, as seems to be necessary on Kitcher's approach. I think we can now also see that such motives are unnecessary in accounting for what Pierre Bourdieu calls the 'social arbitrariness' of 'the scientific field,'[57] in particular, at least insofar as we take it, as we should, as a field in which complexity, in our sense, is rife. The knowledge generated and evaluated within any complex field of enquiry such as the advanced sciences will unavoidably be arbitrary at least in the sense that its development is a path-dependent one. But there needn't, as Bourdieu seems, unlike Kitcher, to acknowledge, be anything sordid about this. The unavoidable arbitrariness of course creates, in effect, a field of possibilities in which sordid motives *might* function, but the arbitrariness is inherent in this field *per se*; it does not arise because or only to the extent that sordid motives are actually in play. Even the saints and heroes of disciplined enquiry are going to generate socially arbitrary, or rather epistemically arbitrary outcomes. There is no alternative.

Of those who may have thought otherwise, perhaps the most articulate was Michael Polanyi who, in his writings on science, specifically posited a kind of hidden hand mechanism that we might associate with the equilibrium approach to explaining and legitimating a division of cognitive labor. He said:[58]

> [S]elf-coordination of independent initiatives leads to a joint result which is unpremeditated by any of those who bring it about. Their coordination is guided by 'an invisible hand' towards the joint discovery of a hidden system of things. Since its end-result is unknown, this kind of cooperation can only advance in stages, and the total performance will be the best possible if each consecutive step is decided upon by someone most competent to do so. We may imagine this condition to be fulfilled for the fitting together of a jig-saw puzzle if each helper watches out for any new opportunities arising along a particular section of the so far completed patch of the puzzle, and also keeps an eye on a particular lot of pieces, so as to fit them in wherever a change presents itself.... We may affirm that the pursuit of science by independent self-coordinated initiatives assures the most efficient possible organization of scientific progress.

Actually, there is some ambiguity here and I should be careful in my reading of Polanyi. Strictly speaking, Polanyi mentions '*a* hidden order of things,' rather than, as might imply *infallibility* in the hidden hand, '*the* hidden order of things.' Likewise, he claims that the kinds of mechanisms we are considering 'the most efficient possible organization of scientific progress,' not, again, that they are infallible in uncovering 'the hidden order of things.' Certainly, there is nothing in the results I've reported that would contradict his efficiency claim. The combination of diversity and division of labor may well be more efficient than other approaches in 'generating a progressive sequence of consensus practices.' But that is a matter for Section 9.3.1, and, in any event, even the most efficient mechanism will, it now seems clear, be a fallible, path-dependent one.

7.4 A liberal approach?

I have, at various points, mentioned, and stressed the significance of, the possibility that both the products of enquiry and the standards by which they are assessed might be made up as we go.[59] The diction of management science discussions of division of labor and complexity has a tendency to obscure this possibility, though it should also be acknowledged that some discussants, e.g. Loasby, have certainly been aware of it. In any event, it is understandable to overlook or at least downplay the 'making it up as we go' aspect of enquiry given the models which have been used to facilitate rigorous discussion. You can't simulate enquiries across a multidimensional epistatically structured problem space if you haven't already constructed it. Nevertheless, I think that this diction has not only made it harder to see that the reality of complexity and evaluation is even more fraught than the models; I think it has also made it more difficult to appreciate another way of approaching the issue of enquiry in the face of complexity.

First of all, let's observe that, if we are making up the problem space as we go, both in its technical features and in our evaluations of its service characteristics, we don't really have a basis for imposing a form of competition between solutions that might eliminate some enquiry strategies that might, in some rather less precise sense than previously, have been meritorious. After all, the point of a competition (not the only point, of course) is to identify an approach which is best, or at least to winnow out the better from the worse strategies for enquiry. And that makes sense only if what is best *is* defined independently of the competitive apparatus itself. But if what is best isn't even well-defined at all—because it is always being made up as we go—then the utility of the competitive apparatus itself must be in doubt.[60] What is it that design teams would be competing about if they are making up both the terrain and the navigational instruments (the values) that they are using to identify a pathway to traverse? I do not think there is any obvious answer to this question that should compel a competition, conceived as such—i.e. as identifying a winner according to predetermined rules.

What does this imply? It might imply that we should be looking for another way of understanding what is involved in dealing with complexity, a way that takes more seriously than either the modularity approach or the population approach the fact that both the objects and the standards of enquiry are being made up as we go.

What does such an approach look like? It is surprisingly familiar. Once we recognize that both objects and standards are being made up as we go, we can see, clearly, that there is nothing interesting to be said about how the space *should* be explored, except to say that it should be explored (as it is made up) in the various ways in which various enquiry teams think best. We should, in other words, devolve decision-making about enquiry to the enquiry teams and let them get on with it. We should, in particular, loosen, rather than tightening, the competitive pressures, at least insofar as we are persuaded by the toy-world simulations of Marengo and his colleagues. But, more importantly, for those simulations themselves embody the false ontology of the preexisting space, we should accept that *there are no independent standards in terms of which it would make sense to standardize enquiry when complexity is present.*

We need, in other words, a *liberal* solution to the problem of enquiry in complex environments. Each team will construct and traverse that region of the space which they find interesting. Some will feel that their construction and exploration have been rewarded and some will not. These judgments are always local and parochial however. And that is that.

7.5 A republican approach

Terminological issues should not too much delay us here, but I do need to say that the 'republican' approach to be described now is not, perhaps, exactly in line with my discussion of republicanism in Section 5.3, though there are some obvious resonances.

One thing which the population approach might have, but which it also disguises, is not very evident either in a certain kind of liberal community. The missing or underemphasized ingredient, in both cases, is communication.

Suppose, as throughout this chapter, that we are concerned with a distribution of enquirers across a complex space of possibilities. (It will matter whether they are 'making it up as they go,' but let's ignore that possibility for the moment—the space, we will assume, is well and antecedently defined.)

Of course, as we have already seen, complexity means that the space is 'rugged' and, accordingly, that certain kinds of enquiry strategies across the space will not always be effective. (See Section 7.1.3.) For example, myopic enquiry may lead to lock-in at sub-optimal local maxima.

All this is reasonably straightforward. What may also be obvious is how, given the ruggedness of such a space of possibilities, we might nevertheless constitute a *community* of enquiry to ensure that the highest peak was indeed selected by exploration of this space.

How could we do this? Actually, the answer is almost naïve enough not to be noticeable. A single individual, exploring a 'rugged' space under pressures of time and other resources, might well fail to discover the *global* 'peak' on this space. For example, if he is exploring the space depicted below under pressures of time and other resources, he might well conclude, having reached it, that the point marked A is a global maximum on this space.

After all, when he proceeds in any direction from A, he finds himself descending rather than ascending. This sets the problem. The solution is obvious within a 'republican' framework. It has two main elements: (1) We need a multiplicity of different enquirers, as with both the population and the liberal models, each exploring a different 'patch' of the terrain; (2) We need communication among the enquirers and, especially, we need them to have a sufficiently common standard of assessment,[61] and a sufficiently accurate technique of measurement, so that, when each of them reaches her local peak, she is able, by making

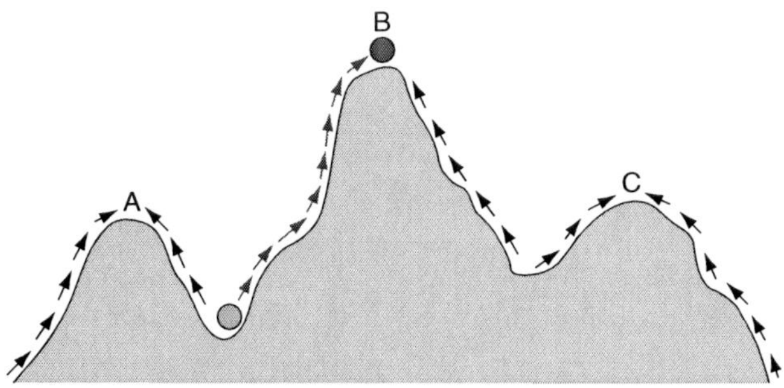

Figure 7.3 A rugged fitness landscape

and communicating measurements, to determine, through an iterated conversation with her fellows, whether her peak is a global peak. (Do my two qualifications probably rule out the applicability of this model when the space itself is indeed being 'made up as we go'? Perhaps not; more on this later.)

To make this more concrete, imagine someone starting from the valley between B and C and heading toward B—she does so, of course, by making such technical improvements to the object (say a theory) she is considering as will improve that object's overall value. When she reaches her local optimum, she can, let's imagine, announce to others, e.g. at points A and C, how high on the contours she has actually gotten. On the assumptions I have made, about accurate measuring techniques and commensurability, all of the summiteers should quickly realize that she who is standing at B is at the highest point on this part of the space of enquiry and, accordingly, should recognize that the option which is represented at this point is the best of those available.[62]

None of this is strictly incompatible, by the way and importantly, with my point about us, typically, making things up as we go. For example, if the participants in this 'republic of enquiry' happen to, or have been so socialized that they do in fact, share certain standards and theoretical assumptions, then they will be making things up as they go in sufficiently similar ways so that republican communication *can* function effectively in their circumstances. And this, of course, is precisely the function of the paradigm, of the negative heuristic and the like. (See Section 6.2.) Of course, the paradigm functions, in this case, analogously with the 'design rules' in a modular system and, accordingly, it precludes exhaustive enquiry over the space, or, rather, an exhaustively searching project of making it up as we go. That the republican approach facilitates a more effective process does not mean that it entirely obviates the path-dependence of our activities.

In any event, such a republican approach is, with an important proviso, 'the logic of patches,' as Stuart Kauffman calls it. He says:[63]

> Have we evolved some other procedure that works well? I suspect we have, and call it by a variety of names, from federalism to profit centers to restructuring to checks and balances to political action committees. Here I'll call it patches.... The basic idea of the patch procedure is simple: take a hard, conflict-laden task in which many parts interact, and divide it into a quilt of nonoverlapping patches. Try to optimize within each patch.

(The proviso is that Kauffman develops a model in which what one enquirer discovers changes the contours of the landscape. We needn't add this element, though it is certainly an interesting one and brings the model back a bit in the direction of my own 'making it up as we go' approach.)

7.6 Interim summary

It is very important to place this chapter in relation to what has preceded it and what will follow it.

First of all, this chapter looks, as I've already said, at something that lies, if you will, at right angles to the problem which has dominated discussion in Chapters 4–6. In those chapters, we were concerned with how we might identify and then overcome the difficulties which social comparison pressures might create for the free exchange of information, for mutual benefit, among the members of a community of enquiry. In this chapter, however, we have not been concerned with either the difficulties or the means necessary and/or sufficient to overcome them. Rather, we have been concerned with a difficulty for communities of enquiry that arises, not because of people's inadequacies as communicators, but, rather, because of the complexity of the 'objective problem situation,' as Popperians used to put it. A great many problems which communities of enquiry orient toward are complicated and, accordingly, are difficult to understand adequately, let alone solve. Division of labor suggests itself as one response to this difficulty, but, as we have seen, division of labor is not a fully adequate response. Even if we make very optimistic assumptions about the ways in which individuals communicate within and across the division of labor within a community, we find that these individuals often will not be able to address their difficult and complex problems effectively. Not because *they* are inadequate, but, rather, because the problems are too hard for merely finite beings to solve in a convincing manner.

This is my first point, to orient us: Difficulties in identifying the correct division of labor exist (or don't) independently of difficulties in overcoming social comparison pressures in order to achieve a sharing of information and hence an assembly bonus. Even if we do communicate effectively, what we have to say to one another will, unavoidably, bear the marks of the paths we have taken to arrive at what we know or believe.

My second point is this. These difficulties put under some pressure some ideas I've already introduced. In particular, I have claimed,

Section 6.8, that communities of enquiry will flourish where they are well matched to the problems they address. There will, I have suggested, be a virtuous cycle of enquiry in such circumstances—individuals tackle problems, make progress, feel rewarded for making progress, and consequently stick with these problems to provide even better solutions. But what if, as I mentioned earlier (Section 7.3.1), the problems are too hard? This is a matter I address, especially in Section 8.4, in the next chapter, where I will also consider some other devices, aside from the division of labor, which we might use as a basis for an effective and self-confident community of enquiry.

8
Some Communities of Enquiry

In Chapter 7, I considered division of labor as one basis on which to establish a community of enquiry and examined the ways in which the complexity of the problems such a community might work on might prevent it from finding optimal solutions to those problems, even when there was good facilitation of the sort of collaboration that is necessary to overcome social comparison inhibitions in collective enterprises.

In this Chapter, I have two main projects.

First of all, I will consider some alternatives to or articulations of the general model of division of labor. In particular, I will consider

- Information markets, and
- The 'federal' model of collective enquiry

Secondly, however, I will consider, via Environmental looping, why the difficulties identified in Chapter 7 need not be *additionally* inhibiting to successful collective enquiry. So, while we have, potentially, both subjective and objective impediments to effective enquiry, the very existence of the objective impediments may serve to 'tune' our communities of enquiry to work effectively in overcoming the subjective impediments.

8.1 Information markets

Novelty premiums and asset ownership provide two essential elements in terms of which to position the independently developed idea of an 'information market,' as a mechanism, in particular, that will overcome some of the social comparison and persuasive arguments influences which can impede the realization, in a community of enquiry, of otherwise available assembly bonuses.

It will be helpful to begin, I think, with some observations by Friedrich Hayek, whose paper 'The Use of Knowledge in Society'[1] provides a crucial intellectual underpinning for the idea of an information market, but, more importantly, indicates what sorts of information can in fact be extracted from such a market.

One of the things which Hayek draws our attention to is the dispersal of concrete knowledge among the members of an interacting population. Someone at some specific social location might have information relevant to the use of or potential market for some commodity or product that is, as it were, 'site-specific,' in the sense that people at other social locations (which need not be distant geographically) would not necessarily have that information about the item's uses. Depending on what the site-specific information is, the individual in question may come to believe that it is worth his while acquiring some of the commodity or product. (Perhaps he thinks of a new use for it for which he predicts good market demand.) Accordingly, he will enter the market for it, and, other things being equal, his doing so will have a tendency to increase the price for that item. (His demand is extra and unless supply has also coincidentally increased, the 'law of supply and demand' dictates a price increase.) And this increase in price is, on Hayek's account, a signal to the community more generally that the item in question is 'in play' when it comes to its uses and substitutions.

The crucial point is: The price signal does not convey the local knowledge that our imagined agent had and which motivated him to enter the market. That may well remain private. As Hayek says,[2] 'There is no need for the great majority of them [agents] even to know where the more urgent need has arisen, or in favour of what other needs they ought to husband the supply.' All that is conveyed by the market signal is that the item is more valuable, for some reason, than was previously thought (collectively, by the market participants).[3] This signal will be enough for people to consider shifting to cheaper substitute factors of production or, if producers of the item, to increase production. Of course, knowing that there is now more demand for the item (an inference from its increased price), some participants in the market may well try to discover why and, if successful, will uncover the information which led our original entrepreneur to enter the market in the first place.

From the point of view of our own interests, the crucial points about such market behavior are these. (1) An individual who possesses important information has an incentive, in the form of potential profits, to use this information. (2) Using the information is not tantamount, however, to announcing it to others because its use consists merely in entering a market for an item (without its being manifest, from that very fact, what the individual's reasons or motives are for doing so). (3) On the other hand, using the information signals others that the item is 'in play' and they may well discover, if not from the individual himself then from their own enquiries, the information that motivated the entrepreneur. (4) Indeed, in many cases, the individual with site-specific potentially valuable information may have a motive, a straightforward economic motive, for keeping the site-specific information to himself. If others acquire this specific information, then they too may enter the market for the item and, accordingly, drive its price higher still, something the original entrepreneur would rather not happen, for the obvious reason that it will cost him more to benefit from his site-specific information. While individuals have no motive actually to reveal the information on which their market interest is based, they do have an incentive to use this information that, on standard assumptions about market behavior, is strong enough to overcome 'social comparison' pressures—they would rather profit from exploiting a new opportunity than simply reproduce existing patterns of utilization.

Despite recent enthusiasm for so-called information markets,[4] it is not clear that these markets actually provide the right sort of framework for securing the realization of assembly bonuses (see Section 2.2) in the face of hidden profiles (see Chapter 4). (And this of course is my problem, whatever the problems are that enthusiasts of the information market might want to consider.) To see this, we reason as follows.

An information market works by getting individuals to trade in shares which are redeemable when a certain proposition is true. The price at which they trade (as a fraction of their redemption value) is an expression of the community of enquiry's estimate of the probability that the proposition is true. Tom Bell provides an example.[5]

> By way of example, consider the 'SLvl' claim traded on the Foresight Exchange, a prominent play-money prediction market. That claim reads, 'By 2030, the greenhouse effect and other causes will have raised the average world sea level by 1 meter from its 1994 level.' The Foresight Exchange facilitates trade in SLvl coupons, each worth one 'FX-buck' in the event the SLvl claim proves true. Since trading opened in 1994, the price of SLvl coupons has fluctuated between about .1 and .5 FX-bucks apiece, indicating how traders' opinions about global warming have changed. For more than a year, the price has hovered near .2 FX-bucks/coupon, demonstrating a consensus view the SLvl claim has only about a 20% likelihood of coming true.

As Bell says,[6] '[b]ecause prediction markets reward correct predictions, they encourage traders to develop truthful beliefs.' We might add: They also encourage traders to express these beliefs through their behavior as they bid for shares. So, for example, someone who had information that more or less conclusively established the truth of the proposition in question would have an incentive to 'bid up' the share price by buying more and more units.

Notwithstanding this crucial fact, and it *is* crucial, it is not clear that information markets really deliver the goods in the case of hidden profiles. Recall that, when there are hidden profiles, A possesses some information bearing on the truth of a proposition and B does too and, together, the information which they have is enough to decide on the truth of this proposition. Recall too that, in some cases, without the combination of the information separately held, neither A nor B has, individually, enough information or the right information to affirm the truth of the proposition in question. Accordingly, in these cases, neither A nor B is likely to bid up the price of shares in the proposition's truth and, hence, price signals are unlikely to function to increase credence in the proposition. In short, there is nothing in the mechanics of the market itself that facilitates A and B combining their information or indeed which gives them an incentive to declare what it is that they know (as opposed to using their knowledge as a basis for market behavior). Of course, as with an ordinary market, when one participant sees that the behavior of others drives prices up, she may well wonder why and seek to find out. But, as with ordinary markets, the costs of finding out may well exceed the benefits and the information may be quite obscure and local. (It might be 'sticky information,' for example. See Section 4.9.) And, in any event, with hidden profiles, there need be no price signal at all and, hence, no curiosity generated about the proposition at issue.[7]

Indeed, in the case of some hidden profiles, there might well be price signals, but they might well create what Sunstein calls cascades of *mis*information.[8] Consider the following scenario. A and B are making judgments about options P and Q. A and B share three pieces of Q-favorable information. Between them, A and B have a total of four (distinct) pieces of P-favorable information, with A having two and B having two. In this situation, each of A and B (without consultation that reveals the hidden profile), actually has reason to bid for Q shares in an information market and, since each bids, each has reason to believe that her own bid was sensible—it is 'confirmed' by the other's bid—and, accordingly, we can expect the 'credence' of Q to rise in such a market, even though it is inferior, as an option, to P.

Notice, furthermore, that, as Sunstein points out,[9] information markets work well even in providing individuals with incentives to act on their information only when certain conditions obtain (which can, of course, themselves be engineered, but which must not be taken as given). Among these conditions are that there be

- a reasonable amount of information dispersed among the individuals constituting the market;
- limited 'reflexivity effects' such as those considered at Section 3.1.4;
- some independent standard of correctness against which individuals making up the market can, at least *ex post*, evaluate their initial positions;
- a reasonable number of participants in the market.

It would appear, then, that information markets are not necessarily a panacea in the case of complex communities of enquiry subject to hidden profiles. They do provide an incentive to use site-specific information, but they do not necessarily provide a context in which that information can be expressed in a fashion which delivers the desired assembly bonus. More familiar mechanisms, such as novelty premiums and asset ownership may serve that function more effectively at least with complex forms of knowledge.

8.2 The 'federal' model of collective enquiry

Some analogies with political institutions may provide an interesting basis on which to summarize some of what we have been considering in the previous chapter. I have in mind, as I did on an earlier occasion,[10] a federal model in which a community of enquiry is differentiated along two axes—levels and jurisdictions, as follows, with some additional refinements which carry the politics/enquiry analogy even further.

8.2.1 Levels and jurisdictions

A jurisdiction consists of a deliberative body and, at every level except the highest (if there is a highest), there is more than one such deliberative body, each acting independently of others at the same level. (The independence requirement may be dispensable.) Hence, at any given level, a deliberative body in one jurisdiction may decide an issue differently than a similar such body in another, parallel jurisdiction. (This is, so far, just a generalization, in effect, of the 'population approach'—see Section 7.3.2.) When this happens, it ensures that, if and when the issue is considered at some higher level, those who consider it there will have two different kinds of information that may not have been available to the original deliberative bodies. First of all, they will have the reasoning that led, in each jurisdiction, to the conclusion reached in that jurisdiction. Secondly, however, they may also sometimes have *ex post* evidence about what happens when a conclusion reached in a given jurisdiction is made the basis for further activities in that jurisdiction. And this second kind of information shows two conspicuous features. First of all, it was not available to either of the lower level deliberative bodies, since it becomes available only as a result of their 'putting their money where their mouths are'—i.e. from their investing in the conclusions which they reach. Secondly, precisely for this reason, this information provides a richer basis for deliberation about the matter in question than was available to either of the lower level deliberative bodies.

Let me explain. Suppose that we have two 'teams' working under the influence of the same paradigmatic approach to enquiry (see Section 6.2.1). Each has interpreted their common paradigm in a distinctive way. Perhaps, both between and within the teams, there is also a certain diversity in the ways in which the evaluative standards are interpreted and weighted that will be used, ultimately, to assess the processes and products of enquiry. In this case, we can expect that the different teams will 'make up as they go' (see Section 7.4) quite different approaches to enquiry. Not only will the products (e.g. findings) differ, but the ways in which these findings were arrived at will differ too—they will look in different places for data, reason in different ways, apply common evaluative standards in distinctive ways, and so on. Whatever conclusions the two teams reach, they may, in due course, try to implement them in various ways. For example, they might try to standardize some of the technology or methodology that they used. Or they might try to engineer some product that reflects their more theoretical discoveries. In any case, the two teams will engender different outcomes from their processes of enquiry.

It is at this point that the idea of 'levels' becomes especially relevant. (Of course, it is important not to artificially separate aspects of enquiry that may well be mixed up together in most actual situations.) Whether it be the editor of a journal or, indeed, members of these two teams themselves, someone will, when seeing that the teams have engendered different outcomes, be interested to discover what the relations are of these outcomes to each other, to the objectives of the paradigm's adherents, and so on. (Which one of them is better? is a rather obvious concern that might be quite widespread among and even outside the community of enquiry.) When someone comes to deliberate about this issue, however, she will have both more and different kinds of information than either of the teams had when they were engaged in their own 'first-order' deliberations. Such an 'adjudicator' will know, first of all, that there are two seemingly different conclusions one can come to from a similar starting-point. She will know about the different pathways which the two teams have explored in coming to the conclusions they came to. She will know how and with what success the seemingly different conclusions can be implemented (or not) in technologies, predictions, and the like. All of this is information that the adjudicator can easily enough acquire that is unlikely to have been available to the original enquirers. We can, at this point, draw the veil away from the 'hidden profile' that may well have existed when the two teams were working in parallel. The adjudicator can, in other words, see what both teams saw, where each of them could see only what could be seen from within the sphere of enquiry that each was engaged in evolving. And whereas each team might have been subject to social comparison pressures, provided there is good reporting of their work, the adjudicator can bypass these pressures and the inhibitions which they engender.

It is important, I think, not to exaggerate the degree to which all inhibitions can be overcome or all complexities avoided through the use of such a federal model. It is not, of course, that the adjudicator (who is rarely, actually, a singleton but usually herself a collective, subject to all the usual infirmities) is simply immune to the sorts of pressures which inhibit the realization of assembly bonuses. It is, rather, that she is better situated or rather that anyone is better situated when they are playing the *role* of adjudicator, than are the ordinary participants in a community of enquiry... or, rather, than such participants are when they are playing a different role—the role of engaged enquiry (rather than reflective adjudication).

Although I draw terminology and my sketchy outline from the political realm, there are obvious analogies to communities of enquiry such as those we find in science or in the professions or, indeed, in organizations such as firms and bureaucracies. Perhaps the jurisdictions correspond to different teams or departments within a larger structure. Perhaps the levels correspond to hierarchies within that structure. This model captures, I think, James Bohman's idea[11] that 'the institutional design [of a community of enquiry] ought to take a distributive rather than a unitary form,' 'making the kind of opportunity for publics to influence decisions possible at multiple locations.'

8.2.2 Representation

I've already mentioned another important feature of a 'federal' model of enquiry, namely, the way in which individual members of a community of enquiry may belong to particular research 'teams' within the larger community structure and hence may 'represent' those teams in the larger deliberative fora in which they play a role. This is the 'multiple accountabilities' of Section 5.1 and it accounts, in some measure, for the way in which a community is able, despite social comparison pressures and other impediments to the achievement of assembly bonuses, to get a full range of opinion represented, and vigorously represented, within the community.

8.2.3 Checks and balances

A conspicuous feature of specifically American federalism is the theory of checks and balances, whereby any given role within the system is, at least potentially, checked and/or balanced by other roles, with the result that the kinds of dysfunctional cascades that Cass Sunstein warns against are, in principle anyway, themselves checked. Let's consider how this might work.

As James Madison pointed out,[12] the willingness of various parties to check and/or balance the points of view of other parties depends on some durable difference between those parties. It is not enough, in his view, that parties are 'accidentally' or 'coincidentally' of different points of view. For, in that case, they might as easily be of the same point of view. And this will certainly be worrisome in the context of the Kuhnian paradigm, with buy-in to its agenda more or less a condition of legitimacy within a given community of enquiry.

Of course, we already have the answer, and a sufficiently Madisonian one. Indeed, there are two main elements.

In the first place, what individual members of a community of enquiry have and must have consensus about are *interpretanda*, not *interpretata*. Their consensus is a shallow one that is consistent with their being, effectively, in disagreement on such specificities as are actually engaged when deliberation occurs. (See Section 6.1.5.) Secondly, as we have already outlined (Section 6.7), members of a community of enquiry will differ, 'non-coincidentally,' from one another because, for example, of their different times of entry into the community, their different seniority, their different primary references groups, and so on. Accordingly, the various members of a community of enquiry will check and balance one another in an appropriate way.

For example, someone with much accumulated social capital, based, say, on her pioneering work in a particular area of study, may choose to invest that capital to champion some new development (and may join in this with very junior members of the community who see this as their best path to the accumulation of substantial social capital). Structural features of her position ensure that she has an adequate basis, in her own reward structure, for sponsoring new developments.

All this confirms, perhaps, the value which Steve Fuller saw in analogizing between political philosophy and epistemology. He said,[13] and I have long believed myself, that epistemology 'is nothing other than the application of political philosophy' to the community of enquiry.

8.3 Communication

Of course, as already indicated, republican (Section 7.4), market (Section 8.1), and federal (Section 8.2) approaches to collectivized enquiry depend heavily on communication. This is important precisely because or insofar as we believe that there are impediments to effective communication—e.g. the reluctances associated with social comparison pressures.

Thanks to Hayek, we understand how centrally communication is built into the market approach in the special, and restricted, sense that prices act as signals that may enable dispersed participants to coordinate their activities. Of course, as I have already indicated, this is all that prices do—they signal that some factor is 'in play'; they do not otherwise inform the market or its participants, and, indeed, each possessor of site-specific, market-relevant information has a motive to conceal this information insofar as she is able.

With the federal approach, there is, again, a direct appeal to the efficacy of communication. Whenever lower level processes and products are under review, their availability to scrutiny depends directly on communications about these processes and products to higher levels of the hierarchy. Again, there is likely to be some disparity between what it would be useful to have communicated and what is actually typically communicated. Certainly, we know that 'gate keepers' can act as filters,[14] reflecting their own agendas, and, accordingly, that some of the diversity that should be evident to the higher level adjudicator is not always transmitted from lower levels.

The communication requirements and potentialities are more complicated with the liberal approach. One reason for this is that this approach recognizes the possibility, though it does not of course insist on the inevitability, of the 'different worlds' inhabited by participants belonging to the same community of enquiry but interpreting its paradigm, its rules, and its underpinning values in different, perhaps even starkly different, ways.[15] With the liberal approach, we therefore have the possibility both that communication will not produce an assembly bonus (because you can't add together 'incommensurables' to get a rational sum[16]) and that communication is likely, in any event, to be fraught because of multiple possibilities for misunderstanding. Although Kuhn has specifically said of pre- and inter-paradigmatic periods of 'scientific' enquiry that they involve participants in 'talking past one another,' this can be true, though not as easily noticed, even when participants work in the same tradition of enquiry. (This is one case, perhaps, where we extend, rather than simply interpreting, Kuhn.)

For example:

1. Two teams may interpret a common paradigm differently and hence take different things for granted or treat different parts of the developing project as fixed (protected by the negative heuristic, in Lakatos's terms).
2. Two teams may interpret a common paradigm differently and hence take different pathways forward from a common starting-point, e.g. because they interpreted differently the directives of the positive heuristic (in Lakatos's terminology).
3. Two teams may differently interpret and/or weight the common standards of evaluation and hence arrive at different judgments about which of two or more options for choice is the better.

Of course, there is an 'upside' to these differences, as I have already emphasized. What I am concerned with here is the 'downside' or at least the downside risk. While the upside is that we get a diversity of approaches, the downside, especially to shallow *consensus*, is that we get diversity without the participants always being aware of it. Sometimes, people think that they share more than they do and hence don't draw the right conclusions from what their interlocutors say.

Consider Case 1 above. If a member of the A team takes it for granted, on the basis of a shallow consensus with a B team member, that they are both working on the same project, then he may not 'hear' her properly when what she says implies (or states explicitly) that she does not share with him some particular assumption. Accordingly, there may never be the sort of 'expectancy disconfirmation' (Section 3.6) which might otherwise arise. Nor will there be an uncovering of a hidden profile that is constituted, in part, by whatever it is that she, B, believes (when it is combined with what he, A, believes). We have diversity and we have the potential to exploit it, but the shallow consensus may mask its very existence. We never put together the A approach with the B approach because or insofar as we do not realize that they are different (and potentially complementary).

In Case 2, we are more likely to encounter a different sort of communication difficulty. In particular, the B team member, seeing that A team members are exploring different parts of a presumptively shared domain of enquiry, may not understand that the A team's enquiries remain engaged with her own team's activities. She may think, rather, that the A team is now engaged in a different sort of enquiry from her own and, accordingly, may come to believe that there is no need any longer to attend to what the A team says or does. We have diversity and we have the potential to exploit it, but the shallow dissensus may mask its very existence. We never put together the A approach with the B approach because or insofar as we do not realize that they are complementary even though they are different.

In many situations, Case 3 is the most perplexing of all. Again, the shallow consensus may mask underlying differences in dealing with common *interpretanda*. In this case, if A and B differ in their evaluations of some particular articulation of their shared paradigm, one or both of them may conclude from this difference simply that the other is mistaken (perhaps even culpably mistaken). There is, as it were, nothing in their situation that calls attention to the source and nature of their differences in judgments and they may all too readily dismiss one another as incompetent... and hence, potentially, as someone with whom it is not necessary to engage. Again, as in the previous cases, we have the potential to exploit diversity, but the possibility that this potential will not always be realized. (Certainly, examples of all these cases will come readily to mind.)

Lorraine Daston provides a valuable discussion.[17]

> For all the cosy *Gemeinschaft* associations of the term 'scientific community,' the actual relationships that welded it together were increasingly narrow and formal.

> But welded together it was, not only by invisible girders that stretched across national and linguistic boundaries in the form of international journals, commissions, and congresses but also by the filaments that crisscrossed levels of skill, status, and training within and among laboratories and observing stations. Articles circulated across oceans and continents, measurements were exchanged, observations tallied, instruments calibrated, units and categories standardized.... *There was nothing inevitable about communicative science*; it required hard work at every juncture: new instruments and new methods of data analysis were a precondition for amalgamating measurements made by far-flung observers; international commissions met and wrangled over the standards and definitions that would make the result of, say, statistical or electrical research comparable; scientific labor had to be divided and disciplined to equalize differences of skill and training....
>
> It would be an exaggeration, but not a distortion, to claim that it was scientific communication that was the precondition for the uniformity of nature rather than the reverse.

The communication from which we can leverage the benefits of a dispersed collectivization of enquiry is itself a notable achievement, rather than a given. When what needs to be communicated needs to be 'calibrated,' the work of calibration, as Theodore Porter has shown and as I have argued on an earlier occasion, is something that, as Daston puts it, requires 'hard work at every juncture.'[18] This hard work goes hand in hand with as it facilitates the division of labor yielding assembly bonuses, or perhaps we should say *dis*assembly bonuses—the bonuses which result when we disassemble (into 'patches') a complex problem into elements which boundedly rational individuals can undertake individually and which a boundedly rational community can use to triangulate a global solution.

8.4 Environmental looping

In Section 6.7, I showed how, on reasonable assumptions, different individuals might accumulate cultural 'capital' in different ways, and, in particular, by displaying, in their stance toward innovation and conservation (the 'essential tension' of Chapter 2), different stances which reflect their own positions in and times of entry into the community of enquiry to which they belong. Similarly, in Section 6.8, I made the in some ways rather banal observation, following Moreland and Levine,[19] that successful performance on collective tasks enhances the capacity of a group to perform tasks collectively, and, hence, their ability to accumulate further cultural 'capital.' If individuals can solve moderately challenging problems, they are confident and some of them will be less averse to risk-taking. They will, accordingly, continue to develop—and yet be ready, as needs be, to abandon—the paradigm of enquiry which guides their work. But whether individuals will be able to solve problems doesn't just depend on their personal characteristics, or, indeed, on the characteristics of their framework for enquiry. It also depends on the nature of the environment in which they work. The degree to which cultural capital can be accumulated in a community of enquiry depends, then, on how well or ill the (inner and outer) *environment* for problem solving supports individual and collective success in problem solving. In adverse circumstances, this supportive feedback loop may not exist. Certainly, the possibility that the landscape being surveyed is a complex one may be of concern in this context.

To develop this point, it will be helpful to begin with some observations of Connolly and Koput. (See Section 7.2.) They are reporting the results of some laboratory simulations of group problem-solving. As they put it:[20]

> These demonstrations present subjects with environments in which hidden variables are strongly coupled, important patterns build slowly, responses are delayed, and feedback loops are perverse—probably realistic representations of real-world complexity, but cognitively overwhelming. Dorner, for example, reports such dysfunctional behaviors as excessive focus on a single problem and/or sequential problem hopping; failure to cope with slow systems response or long-term trends ('last-period focus'); and response rigidity in the face of deteriorating system performance.... Highly dynamic, complexly looped environments are certainly difficult to understand and operate.

As Connolly and Koput quite reasonably point out, dynamic and complexly looped experimental environments are 'probably realistic representations of real-world complexity,' as many enquirers, in many fields, might encounter it. (This is the kind of space of enquiry that we considered in depth in Chapter 7.) Certainly, Winterfeldt and Edwards are trenchant in their implied characterization of the sorts of 'real-world' problems that a community of enquiry might face in science or business or the humanities. They say:[21]

> Textbook examples of decision problems usually come neatly structured, with options, objectives, and uncertainties prespecified. In the experience of most decision analysts, structuring problems and identifying options and objectives are the most difficult parts of most problems. As initially presented, many problems do not offer an opportunity for choice among options. The most common initial condition is a set of often vague values, typically inconsistent with one another, and a sense of perplexity about how to find a sensible course of action that at least partially serves them all. In such cases the first and most important step in structuring must be to translate the problem into a set of alternatives and values, options and objectives. A later important step often is to identify uncertainties about the outcomes of the available options and to think about acts that may help to reduce some of these uncertainties or allow the decision maker to hedge against them.

Similarly, working from a different tradition, Luigi Marengo and his colleagues, already quoted at Sections 3.1.6 and 7.3, agree that many 'real world' problem situations (as a Popperian might call them) are messy in a way that is rarely adequately captured in theoretical writing about deliberation, investigation, and decision-making. As they put it:[22]

> In problems whose solutions involves the exploration of high-dimensional spaces, agents endowed with limited computational capabilities and with a limited knowledge of the interdependencies can explore only a subset, possibly very small, of solutions. Even if we assume that the selection mechanism which selects among alternative solutions works perfectly..., the outcome of selection is bound by the set of solutions produced by the generative mechanism. It may well be the case that optimal or even 'good' solutions will never be generated at all, and thus that they will never be selected by any selection mechanism whatsoever.
>
> The problem here is that strong interdependencies create a large set of local optima in the search space. Marginal contributions rapidly switch from positive to negative values, depending on which value is assumed by other components. As a consequence, the presence of strong interdependencies prevents the possibility of reaching optimal

> solutions by simply adopting an optimal value for each of the components a problem is made of. It is thus possible that, given an n-dimensional problem whose current state is $a_1, \ldots, a_n$ and whose optimal solution is $a^*_1, \ldots, a^*_n$ some or even all of the solutions of the form $a_1, \ldots, a^*_i, \ldots, a_n$ have a worse performance than the currently adopted one. Also note that if each of the a_i's was traded on a competitive market with prices reflecting their revealed productivity, notionally superior resources a^*_i would never be hired as their marginal productivity is negative. As a consequence... it might well be the case that the optimal solution will never be generated and thus never selected.

If these accounts are accurate and relevant, and the coincidence of these findings across different disciplines and underlying ontologies certainly supports the intuition that they are, then this suggests that typical enquirers will exhibit a variety of dysfunctional behaviors. And, indeed, this is what we tend to find, according to Kuhn,[23] in pre- and inter-paradigmatic science: individual enquirers hopping from problem to problem, becoming fixated on intractable issues, and the like, much as Dorner reports. In such circumstances, enquirers, individually and collectively, are going to be, as Moreland and Levine put it, 'consistently unsuccessful at achieving their goals,' and, accordingly, are going to lack confidence and remain risk-averse and mired in an environment in which social comparison processes play a more important role than processes of enquiry aimed at the phenomena under investigation.

All this changes, though, when individuals address such a highly dynamic and complexly looped environment in a systematic way as a temporally protracted collective project in which a simplified model of the environment is gradually, and in Lakatos's terms,[24] progressively articulated to that environment. Let me explain. (This is the 'flip-side' of my comments earlier, in Section 4.3, about the possibly baleful influence of paradigm commitment and it echoes and expands on some of my comments in Section 6.2.1.)

What a paradigm or research program provides for a community is a framework for dealing with 'real-world complexity' by addressing it, as Connolly and Koput put it,[25] through 'local, incremental, conservative adjustments to reduce tension in specific parts of the values/actions/plans network rather than broad, comprehensive, radical optimization.' In other words, such a framework provides, if you like, a plan of action for addressing a complexly looped environment which requires of each individual enquirer merely that she contribute to the gradual and progressive articulation of this framework to that environment at the particular locus in that environment which she occupies. (This should sound a bit like the 'republican' approach outlined in Section 7.5.) Should she be able to make some contribution of this kind, she will be rewarded and honored. If she should not be able to articulate the framework to the environment, she will, as already mentioned (Section 5.8) consider herself and be considered by others to have failed. But she will not, in either case, have to face the complexity and loopiness of the environment head-on, as it were. She is a soldier, not a general; her task is a local, incremental, and conservative one... and, more importantly, everyone is a soldier; there are no generals; and the group's approach to their environment of enquiry is one of gradual attrition across a front as broad as the membership of the group and effective communication among its members will sustain.

It is important to be clear about my claims here. I am *not* claiming that the paradigm or the research program provides a way out of the arbitrariness which I identified (Section 7.1.3) in the exploration, by whatever means, of complexly structured problem situations. What I *am* claiming is that orientation to the paradigm, rather than to the environment that it facilitates the exploration of, keeps individuals oriented to a situation in which they can make progress and, more importantly, can see themselves (and be seen by others) as making progress. What the enquirer working in terms of a paradigm, working on a research program, is doing is articulating the paradigm; she is only secondarily (and only in a mediated way) exploring the underlying problem space. Insofar as she makes the paradigm fit the local phenomena, and vice versa, she will be recognized by her peers and will recognize herself as having done her job. What she will be aware of, with this focus of attention, is local progress; it is not global complexity and the uncertainty that it raises about whether local progress is global progress. The paradigm provides, if you like, a facilitative myopia even when the search is not myopic in a technical sense of Section 7.1.2.

Of course, all this is very allusive. Can I be more precise and put the matter in a less effusive way? I think so.

First of all, and as already considered (see Section 2.1), individual enquirers working in a particular domain of enquiry are likely to, and it is desirable, *ceteris paribus*, that they have a variety of different *evaluative profiles*, or, to put it differently, systems of trade-offs among the numerous values (see Section 6.1) that are relevant to evaluating the findings that their enquiries produce.

Secondly, because what the enquirers are dealing with when they encounter a particular problem-situation is an *interpretandum*, not an *interpretatum* (see Section 3.5), their various individual differences—in evaluative profile, in risk aversions, and the like—will position them diversely in relation to the various interpretations of this problem-situation that are available.

What this amounts to, then, is that different members of a community of enquiry, deliberating about or investigating one and the same problem-situation, will, in fact, position themselves, in effect, at different starting-points on a 'fitness landscape' covering the potential solutions to the problem which they are addressing.[26] And, because of these different evaluative profiles, they will move differently, from these different starting-points (or even indeed from the same starting-point), across this landscape. The landscape will, accordingly, be much more thoroughly explored by such a community than it could be, *ceteris paribus*, by a solo enquirer.

And, if there is indeed a peak on this landscape—a solution to the problem—which is of more than merely local significance, that has some claims to be a global optimum, then such a community, if they are communicating properly, is more likely, *ceteris paribus*, to find it than, again, a solo enquirer. And, subject to the usual vicissitudes, which I have already extensively canvassed (Chapter 4), and in view of the various facilitating conditions which I have also identified (see Chapter 5, especially the discussions of accountability and the novelty premium), there *will* be adequate communication within this community and, accordingly, a global optimum solution may well be discovered, e.g. through the gradual articulation of a 'dominant' solution (see Sections 6.1 and, more importantly, 6.9).

In other words, and as the history of science itself notably shows, even complex environments can be mastered when they are addressed patiently and collectively. And rewards are available for enquirers working in this way; rewards that sustain their confidence without promoting complacency, and, hence, make it possible for them to trust one another and the framework which guides them, and, accordingly, to reap the assembly bonuses which, in these kinds of environments, have to be realized if enquiry if to succeed.[27]

What emerges from this analysis is, as I've already said, an argument that the ways in which human enquiry are organized trade, in a positive, rather than a negative feedback loop, upon the difficulties of conducting enquiry. Precisely the environmental challenges which might hamper some forms of enquiry in fact support enquiry when it is conducted in certain kinds of ways, as described, for instance, by Kuhn and Lakatos. Conditions which might seem to set limits on enquiry in fact elicit the development of a mode of enquiry which makes even these conditions amenable to investigation.

9
Naturalizing the Essential Tension

I have tried to map out some of the key elements in communities of enquiry. I started with Kuhn's essential insight, as I see it, and the founding gesture, again as I see it, of a genuinely collective epistemology—namely, his so-called risk-spreading argument (see Sections 2.1, 6.1.2). I have tried to provide an account of the 'central tension' or, in Kuhn's argot, an 'essential tension' in any community of enquiry that is to be understood within the constraints (and opportunities) of our own boundedness in reason and in other resources (see Section 2.3). In particular, I have tried to identify, first of all, the opportunities for the collectivization of enquiry; secondly, the impediments to the effective collectivization of enquiry, especially in those psychic and social pressures which, *ceteris paribus*, often limit our ability and will to cooperate for mutual advantage; and, thirdly, the facilitators, institutional, motivational, and cultural, that may, when brought into play, sometimes enable us to overcome the impediments to collective enquiry and, thereby, to reap its rewards. Prominent amongst the impediments to collectivization are social comparison pressures, first-mover advantages and accompanying equilibrium traps. Prominent amongst the countervailing facilitators are multiple accountabilities, novelty premiums, and performance, cultural, and environmental feedback loops that can reward individuals for their cooperation and can, indeed, 'tune' that cooperation to highly effective levels. We have, I think, reason to believe that some such balance of these forces pro and con has been achieved, institutionally and culturally, in some of our most familiar institutions, including jurisprudence, science, and the learned professions more generally. These areas are all, of course, ones in which complexity is an issue and we can therefore expect, despite whatever balance between inhibition and facilitation we might have achieved, that processes and outcomes in these areas will be path-dependent. There appears, really, to be very little alterative to that.

On the other hand, notwithstanding the possibility that some human institutions are reasonably well tuned to collective enquiry and deliberation, a number of issues remain, and I will consider them in this, the final chapter of this book. In particular, I will consider five key topics, as follows:

- Path-dependence revisited
- Assessing the community of enquiry
- The secular turn
- The issue of scalability
- Philosophy as cartography

First, however, a summary of the model of enquiry which I have been developing.

9.1 The model summarized

I began with some observations about Thomas Kuhn's great work and, especially, a feature of it that has, in my view, attracted too little attention. I mean, in particular, Kuhn's risk-spreading argument, which only emerged into the literature about Kuhn about 30 years after it was formulated by him. I positioned the idea of risk spreading in relation to another, also under-recognized theme of Kuhn's—namely his concern, from the beginning of his career, with the 'essential tension,' as he called it, between two attitudes or points of view, both of which had to be recognized and indeed institutionalized within any community of enquiry that was going to function efficiently and effectively in its activities.

In subsequent chapters, I have cobbled together, largely from materials not normally cited by theorists of enquiry, a naturalistic account of how the essential tension might be maintained, specifically through a variety of mechanisms and institutional forms and always in the context of certain features, more or less inescapable, of the human condition. Overarching all the more particular elements is, of course, the boundedness of humanity—our inability to survey all that is relevant, to consider all potentially informative points of view, to infer all that might follow from or be suggested by what we do know, and so on. Among the more particular elements, one stands out as of supreme importance, for it is, as we have seen, the one potential impediment to achieving a division of labor, to securing a balance between risk-taking and conservatism—I refer, of course, to social comparison. As much as we might benefit from sharing information and points of view with one another—as much as expectancy disconfirmation might enhance our own individual habits of mind—we are often, in our everyday lives, inhibited from doing so by our 'conformist' psychology, to use an expression of the Fifties (to complement the expression, Getting It Together, which I borrowed from the Sixties).

In the course of my hunting and gathering, I came across a number of different mechanisms which, presumptively (and in some cases transparently) do disinhibit our communications with one another. Perhaps the most important, in particular and exemplary communities of enquiry, are novelty premiums and multiple accountabilities. Certainly, the exploitation, in communities devoted to the human excellences, of the so-called fundamental attribution error, otherwise a rather unattractive and potentially crippling element of the human psyche, is fortuitous. Thus a balance is achieved, a tension is maintained. Because or insofar as participants hold themselves responsible to achieve their goal (and to answer to their fellows), they stand some chance of overcoming the inhibitions which are otherwise quite common in human affairs, and which, when they *are* present, can strangle effective communication and problem-solving.

Some of the elements we have assembled here appear rather equivocally, and that is perhaps informative (e.g. about the inherent ambivalence of our humanity). Paradigms appear as inhibitors of effective collaboration in Section 4.3 and, in Section 6.2, as a crucial element of a culture of enquiry that maintains a proper tension between consensus and diversity, especially as consensus is necessary if communication is to occur while diversity is vital if the communicants are to have anything worthwhile to say to one another.

Equivocality, though probably without ambivalence, also characterizes my approach to the analogies between enquiry more generally and certain political forms more specifically. Certainly, a republican culture of enquiry and a republican approach to the division of labor show two different facets of what is itself a complex political idea. Similarly for the idea of path-dependency, which appears, in Section 4.6, in relation to the equilibrium traps that can inhibit communication and, in Section 9.2, as an inevitable concomitant of our humanity and as a shaper of all our products and projects.

All these points are summed up in Figure 9.1, which, I think, redeems my promise to naturalize the social epistemology which has been

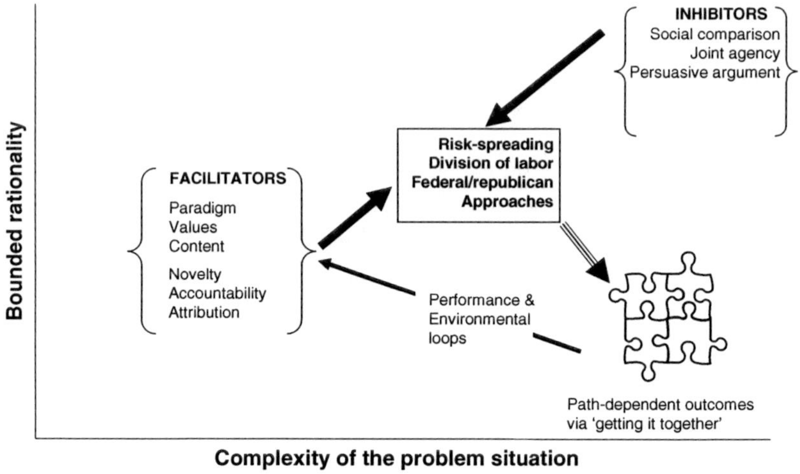

Figure 9.1 Dynamics of a community of enquiry

sketched by others by, especially, embedding abstract ideals of enquiry—share information, divide the labor, replicate one another's work—in specific features of human psychology and of the social circumstances of knowledge work, and in the underlying objective characteristics of the problem situations which enquirers encounter, in particular, and as Chapter 7 shows clearly, the complexity of the issues dealt with by human enquirers, a complexity that is not even relative to their boundedness; it is complexity that wouldn't disappear even were bounds to be relaxed.

There are, then, both objective and subjective impediments to effective enquiry by organized or at least culturally supported communities of enquiry. While division of labor and risk spreading ameliorate these difficulties, they do not overcome them. For these, and other reasons, all our findings, and the processes we employ to derive them, are products of our history, rather than deliverances of a merciful providence. What we know how to do and what we have come to believe reflects the paths we have taken. Of course, as the arch-pragmatist Richard Rorty himself pointed out, our skills and knowledge, if properly disciplined, will also express something about the world (and not just about us). We do not abandon a certain kind of realism when we acknowledge the inescapably path dependent character of our enquiries.

This brings us to . . .

9.2 Path-dependence revisited

As Jesus Zamora Bonilla has put it,[1] *'the evolution of scientific knowledge will be path-dependent.'* I agree, but it is important, nevertheless, neither to exaggerate nor to underestimate the degree to which the deliverances of communities of enquiry are path-dependent, and would have been different, perhaps different in arbitrarily significant ways, had the 'first mover' not made the particular contribution which she did make.

It is important not to exaggerate the path-dependence of human enquiry because, of course, there are various 'corrective mechanisms,' some of which I have been at pains myself to identify and to characterize, which function, precisely, to shift deliberation away from 'generatively entrenched' positions (see Section 4.7). In particular, multiple accountabilities (Section 5.1), novelty premiums (Section 5.4), and a liberal approach to the cognitive division of labor (Section 7.4) across distinct individuals or teams within the larger community—all these mechanisms, supported, in turn, by a republican culture of enquiry (Section 5.3), give us, collectively, some control over, if not the directions taken by our enquiries, then, at least, over the destinations we arrive at.

On the other hand, there is no point pretending that path dependency (Section 4.6), generative entrenchment (Section 4.7) and the institutionalization of dominance (Section 4.8) play no role in the course our enquiries take. As we have already seen, paths to which there would have been alternatives (which might have had better outcomes) do get entrenched precisely through the investments which individuals make in them, serially and in parallel, and precisely because it is easier and potentially more efficacious to contribute to and thus reinforce an already existing method of enquiry than it is to try to attract attention to and support for an entirely new one, promising 'the road less travelled.' This is one reason why, as Kuhn subtly observed, scientific revolutions, which do entail 'changing course' rather than merely 'more of the same,' are so fraught emotionally as well as intellectually; generative entrenchment is often accompanied by 'commitment' and hence is implicated in identity formation and self-esteem. 'Letting go' or 'moving on' or 'getting over it' can be difficult in these circumstances, and while this inertia can serve functions essential to enquiry, as Kuhn might have put it, it can also contribute to entrenchment of what was, after all, only one, and perhaps not even the best, of the available approaches, positions, or theories.

All this of course reinforces some points that Richard Rorty made when, for example, he distinguished[2] between a model in which nature or society insists 'on being described in a certain way, its own way' and a model, on the other hand, where the social or natural phenomena 'can, given a prior agreement on a language game [e.g. paradigm], cause us to hold beliefs [or, more accurately, through falsifications, to abandon them], but ... cannot suggest beliefs for us to hold' or, indeed, point us, unerringly, toward the right sets of beliefs to arrive at (even if we do not start with them). Where there is path-dependency, there is, in whatever our findings turn out to be, a two-fold 'foundation.' On the one hand, and as insisted by the scholars whom Rorty calls 'representationalists,' there are the phenomena our enquiries are about—what they are like (though we can only know this through our enquiries) influences what our findings about them are like, especially in forcing us to reject accounts of their natures that do not enable us to manipulate them or to bring them under a satisfying form of intellectual control (e.g. in developing reproducible results, or 'facts,' as Joseph Rouse calls them[3]). On the other hand, and as insisted by, say, 'constructivists' (though often without due acknowledgment of the first point), our findings depend, of course and unavoidably, on the paradigms, rules, values, heuristics, and other path-dependently developed devices that we use as the tools of our enquiries. As Kuhn insisted, we have no other disciplined way of approaching the world except through these sorts of devices, and, hence, no device-independent understanding of the world.

Could we have arrived at a wholly different scientific or humanistic worldview if we'd followed a different path than the one we did follow—the path initiated by, take your pick, Plato, or Galileo, or Moses? Probably we could have. After all, we know that, in other cultures not until recently too much influenced by the larger 'western tradition,' people did arrive at very different naturalistic and humanistic (and spiritualist) worldviews, no doubt reflecting the generative entrenchment of their own 'first movers.' Does it matter that the path we have trodden is not, in all likelihood, the path, of all those we might have trodden, that leads to some utopian optimum of knowledge and efficacy? Probably not, especially since the very definition of a 'utopian optimum' probably makes no sense, given the plurality of competing standards and approaches and, especially, given that, as I have put it, we are making it up as we go, where the 'it' we are making up includes both the projects we undertake and the standards that we use to evaluate progress in these projects.

Notwithstanding this 'che sera, sera' attitude on my part, there is still, I think, a residual normative question that, armed with a reasonably detailed discussion of the 'essential tension,' we might now approach, however gingerly. Of course, we must be prepared to find that such an enquiry proceeds in a 'pure proceduralist' way, as I have, following Rawls, put it elsewhere.[4]

9.3 Assessing the community of enquiry

One way of thinking about my project is that I have tried, in the preceding pages, to develop a schematic design for communities of enquiry. I have tried, if you will and as my occasional chapter-ending diagrams are meant to suggest, to identify the parallelogram of forces that transects any community of enquiry and, when the forces themselves are balanced, enables that community to function effectively in its environment (and given the various infirmities of its members and of their interactions with one another).

A properly functioning community of enquiry, I have argued, by implication anyway, is one in which assembly bonuses are realized because or to the extent that social comparison pressures and other inhibitors of collective interaction, are counterbalanced by various facilitating conditions, such as novelty premiums, multiple accountabilities, and the like, all this superintended and perhaps even 'tuned' by various feedback loops.

Of course, it cannot be guaranteed that, even when an effective balance has been struck, the community *transcends* its finitude and fallibility. Indeed, we can pretty safely predict that it will not transcend these inevitable limitations on its processes and outcomes. This is so, as the argument of Chapter 7 shows clearly, because of the nature of the problems likely to be addressed by any even moderately interesting community of enquiry. Insofar as its problems are addressed against multiple, interdependent standards of evaluation, a community of enquiry is going to be more than merely casually subject to path-dependencies.

There is, of course, another matter. If we add that both the evaluative standards and the objects of evaluation are being, typically, made up as we, the community, go, we can see, I think, that what I used to call a 'teleological' approach[5] to the evaluation of the community will probably be out of order. To put it crudely, if we do not have antecedently defined standards of evaluation (incorporating a full and definitive schedule of trade-offs when different standards compete with one another), then, as with the concept of justice in Rawls's formulation,[6] we cannot define the success of the community in terms of the outcomes of its processes as measured against these (antecedent) standards. (Simply: the outcomes and the standards 'co-evolve.') And this is so not for any loosely 'post-modernist' or even traditionalist skeptical reasons. This is so, in particular, because evaluation of processes against their 'truth-conduciveness' presupposes that we can tell what 'the truth' is, independently of the processes used to generate candidates for 'the truth.' And, of course, this has been notoriously difficult to argue, in view, for example, of such staples of modernist epistemology as the lack of an 'empirical basis' and the inability to derive 'the truth' in interesting domains from axioms whose own truth is indubitable and so on.[7] We can't, in other words, evaluate processes by considering the truthfulness of the outcomes they generate because we have no means of determining truthfulness in this sense (except, of course, in terms that presuppose the processes themselves).[8]

So how *are* we to define success, if not by evaluating the outcomes of the community's processes—i.e. teleologically?

We are to do so, much as Rawls himself suggests, by evaluating the community's processes directly, against standards which are applicable to the evaluation of just such processes. This is already familiar, perhaps, in the distinction between procedural and substantive justice in relation to the law. To invoke substantive justice is to presume or demonstrate that there is a well-defined understanding of what substantive outcomes justice requires in particular cases and that legal processes are good (or bad) to the degree that their outcomes correspond to or approximate the substantive outcomes required by an antecedently and independently defined notion of justice. But what if 'the law' and our substantive understanding of justice 'co-evolve,' as I have put it? Then 'substantive justice' is not available as an *independent* measure of the success of the law. So what is? How can the law be evaluated with respect to ideas and ideals of justice? It can be evaluated 'procedurally'? There are, we argue, procedural niceties which have to be honored if the way in which the law is elaborated is to be considered just.

Just so with the evaluation of communities of enquiry. If we cannot say of any even moderately interesting community of enquiry that some particular concretely (and independently) defined outcomes are the outcomes which it *should* have produced and, accordingly, that it is a good (or bad) community to the degree to which its outcomes correspond to or approximate these outcomes, ... then we can say, comparing the two communities, that the first did to a certain extent honor the procedural niceties associated with communities of enquiry and, to the degree that the second fell short of this performance, the first is to be preferred to the second. This, anyway, is the approach I will take here.

The basic proposal, then, is this: Communities of enquiry are to be evaluated against certain procedural standards (rather than in terms of their success at producing certain outcomes). All very well and good, but what are the standards? Here, the management theorist Kenneth Knight provides some helpful, though of course again multidimensional, guidance. (This meta-level multidimensionality is what we should expect.[9]) Knight says:[10]

> In analyzing organizations... I have found it helpful to use... six criteria... as a way of distinguishing between the different requirements which organization structures have to satisfy.
>
> 1. *Efficiency*...
> 2. *Control*...
> 3. *Accountability*...
> 4. *Co-ordination*...
> 5. *Adaptation*...
> 6. *Social effectiveness*....

This rich set of standards will bear explication. (Notice, by the way, that all these standards are procedural, rather than outcome-oriented.)

9.3.1 Efficiency

Here's what Knight has to say about efficiency:[11]

> This is a narrower concept than overall effectiveness. It refers specifically to the ratio of outputs to inputs which is the economist's definition of efficiency. Increased efficiency comes from maximising the use of available resources, increasing outputs without increasing cost, or providing a level of production or service at minimum cost.

How does a 'well-balanced' community of enquiry, along the lines we've been sketching, measure up on this score? Certainly, this is a reasonable enough requirement, in the case of communities of enquiry, especially in light of our reiterated discussion of the boundedness of both individual and collective agents in the development of knowledge, understanding, and skill.

To consider the efficiency, in this input/output sense, of various communities of enquiry, it will be helpful to begin precisely where we did begin. There is a tension between the inhibitors and the facilitators of communication. Consider a community in which the inhibitors are not counterbalanced by the facilitators, or, in other words, where social comparison pressures, broadly construed, so inhibit individual members of the community that they fall back, in their deliberations and conversations, on common knowledge, rather than tapping into the local knowledge that each one of them might have. This is clearly inefficient in a straightforward sense: in fact, they have resources which they squander. If A and B were to put their information together, they would realize an assembly bonus which, because they do not, is never realized. A community in which these inhibitors *are* counterbalanced by facilitators such as multiple accountabilities and novelty premiums does not squander inputs in this way. A community of enquiry of the kind I have been describing is, at least minimally, efficient.

We could, of course, consider how well 'tuned' the community has to be to be 'really' efficient. This is an issue which it is hard to talk about in the abstract. Consider, for instance, how, in some environments, and relative to certain kinds of problems, it may be more efficient to limit diversity rather than encouraging it. Notice, furthermore, that, because of path-dependencies in particular, it would be foolish to claim that even a 'well-balanced' community of enquiry is 'Pareto-efficient,' in the sense that no improvements on its processes were possible. After all, even as it makes certain assembly bonuses available, through the disinhibition of diversified communication, it blocks other pathways that might have led to 'better' destinations.

These niceties aside, a community which balances inhibitors and facilitators in some way is clearly more efficient than one in which there is too little facilitation and too much inhibition.

9.3.2 Control

In relation to control, Knight has this to say:[12]

> This is the steering function. Is the organisation so structured as to be able to decide where it is going and take action to get there? The control criterion includes three parts: the ability to set objectives, the ability to monitor their achievement and the ability to take corrective action where necessary.

How does a 'well-balanced' community of enquiry, along the lines we've been sketching, measure up on this score? There are two main points which need to be mentioned here.

First of all, all the devices associated with the specifically Kuhn/ Lakatos model of enquiry—paradigm, positive heuristic, a complexity of shared values—serve a steering function in a well-balanced community of enquiry... while, it must be added, allowing a certain diversity of interpretation that, typically, does not automatically preclude meaningful collaboration. It is the *interpretanda* of iconic achievements, standards, and research agendas that enable members of the community to set objectives, monitor achievements, and take corrective action.

Secondly, and perhaps ironically given this first point, the specific kinds of control devices identified by Kuhn are, in the language of H.L.A. Hart,[13] 'open textured,' requiring interpretation and, hence, permitting a diversity of interpretations and hence spreading risk. But more on this second aspect below.

9.3.3 Accountability

Accountability is crucial, according to Knight, who sketches this desideratum as follows:[14]

> This may be seen as a part of the system for exercising control, but there is a separate point. Hierarchies function by assigning responsibility for tasks to individuals. The ability to hold people accountable for defined task areas and the achievement of specified goals is not just a way of keeping them in line, but constitutes a powerful means of motivating them to exercise discretion constructively and creativity.

How does a 'well-balanced' community of enquiry, along the lines we've been sketching, measure up on this score? Rather well, I would have thought.

Certainly, the fundamental attribution error, as I have argued, is, in a well-balanced community of enquiry, built into the background culture and allocates responsibility for non-performance directly to individuals, thus keeping them up to the mark. More subtly, perhaps, we have the multiple accountabilities—of the enquirer both to her particular primary reference group and to the community as a whole—that disinhibit individuals and thus make it more likely that they will share information with one another, and thus realize available assembly bonuses. Asset ownership also plays a role in securing accountability, especially in providing members of the community with extrinsic motivations to play their roles in the activities and projects of the community.

9.3.4 Coordination

Coordination is particularly important where there is complexity and a division of labor, as Knight recognizes:[15]

> Because all organisations get work done by some form of division of labour, they have to have means of integrating the efforts of groups and individuals towards composite goals. It is necessary to avoid situations in which different parts of the organisation are pulling in different directions, sub-optimising or working at cross-purposes to each other.

How does a 'well-balanced' community of enquiry, along the lines we've been sketching, measure up on this score?

There is quite a lot to say on this matter and, indeed, Chapters 6 and 7 constitute an extended account of these issues, for the *kind* of community of enquiry I have been sketching, and, by implication, recommending as a paradigm for such communities.

First of all, division of labor is crucial, especially in relation to boundedness in the rational capacities of individual enquirers and in relation to the complexity of the problems which communities of enquiry typically address. Having said that, it is, of course, precisely the conundrum to which this book is addressed to indicate, even schematically, how individuals working collectively on a project, and thus exercising their joint agency, are able to do so in a coordinated and yet diversified way. This, I think, is the significance of the notion of a 'shallow consensus' which I have tried to elaborate from less than explicit Kuhnian sources. Individuals are able to coordinate their activities and attitudes because or to the extent that they share a paradigm and a set of evaluative standards. On the other hand, they are able to diversify their enquiries, spread the risks of overcommitment and the like precisely because what they share are, as I have repeatedly put it, *interpretanda*, rather than *interpretata*. They may come to different conclusions, but they address one another as members of a community (and not as unintelligible strangers) because or to the extent that they recognize one another as sharing an, admittedly 'open-textured', set of commitments and ideas.

Also relevant in this regard is the way in which the moral hazards of agency can be balanced by an appropriate system of asset ownership. Certainly, to the degree that individuals might withhold information which they can see some advantage in retaining for their private benefit, they will not effectively coordinate their actions with others, especially if they too are withholding locally-available information.

On the other hand, insofar as an effective regime of asset ownership, insofar as an effective 'economy of esteem' has been implemented, these selfish impulses, destructive of coordination, will be effectively counterbalanced.

Perhaps ironically, given their potential to impede effective communication, the so-called norms of joint agency also function to secure coordination of distributed activities. They give each individual participant the sense that her interests are bound up with those of her collaborators.

9.3.5 Adaptation

Since the circumstances in which knowledge is developed and assessed is constantly changing, the adaptability of any 'knowledge organization' is crucial to its success, as Knight recognizes:[16]

> The organisation's environment does not stand still. An effective management structure has to have the capability to anticipate and respond adaptively to new and changing demands, from its clients, its providers of finance, its employees and the labour market they come from, the community and society at large. To respond quickly enough to new and unexpected requirements it has to be able to recognise them, to solve problems and to innovate.

How does a 'well-balanced' community of enquiry, along the lines we've been sketching, measure up on this score? The key to answering this question is the discussion at Section 6.7, of the endogeneity of a distribution, within a community of enquiry, of risk-aversion. Let me explain.

'[T]he capability to anticipate and respond adaptively to new and changing demands' depends on having, within a community or organization, some participants who are relatively less averse to the risks of new ideas or new approaches, or, alternatively, are relatively more open to the possibilities of new paths or, indeed, to the necessity of striking out in new paths. And the presence of such individuals within a community of enquiry is, as I explained in Section 6.7, a more or less inevitable consequence of some fundamental sociological facts about the community and, in particular, the fact that participants enter the community at different times and, accordingly, have (though not solely for that reason) different accumulations of cultural capital and, accordingly, different attitudes about potentially risky behavior.

Of course, within one environment, there might, whatever these general background facts, be relatively little reward for risky behavior and, in that environment, 'tuning' will probably dampen risky behavior. On the other hand, where there are 'new and unexpected requirements,' there is greater potential for risky behavior to be rewarded and, accordingly, environmental 'tuning' of the community and its members will tend to enhance the place and significance of risk-takers.

9.3.6 Social effectiveness

As we have repeatedly urged, it is crucial, to the success of any institution or practice which is aiming to produce and assess knowledge-claims that its participants be engaged in its activities. Knight already saw this clearly:[17]

> As well as being a machine for performing work, an organisation is a social system. Its structure has to be viable socially as well as technically. This means that it must be able to satisfy its members' needs sufficiently to enlist their commitment to the organisation, and it must structure roles and relationships so as to facilitate co-operation and minimise harmful conflict between members.

How does a 'well-balanced' community of enquiry, along the lines we've been sketching, measure up on this score?

The idea of social effectiveness is really the idea, in my terminology, of 'uptake.' How effectively can an organization or community recruit participants to play the roles that need to be played to realize the objectives of the organization? Several mechanisms are, I think, at work in well-functioning communities of enquiry (though, of course, this is a contingent fact, not an historical inevitability).

First of all, the general line of reasoning, developed in Section 3.1, would be familiar to most potential participants in a community of enquiry—they will realize that their chances of understanding the phenomena they are interested in are enhanced by their participation in a collective approach to securing such an understanding.

Secondly, the novelty premium and asset ownership are powerful inducements and contribute to an economy of esteem in which individuals can hope, through faithful participation in the work of the community, to secure social capital.

Finally, it is crucial that we understand, in relation to this particular desideratum, that effectively functioning communities need to secure different kinds of uptake. For example, the community depends on some balance between risk-takers and risk-avoiders. It will need, and as we have shown it can produce, a diversity of personnel with respect to this dimension.

9.4 The secular turn

Steve Fuller provides an important context in which to understand my enquiry on this occasion. He says:[18]

> In contemporary philosophical parlance, 'naturalism' is the view that normative positions must be constrained by what is empirically known about human beings. Naturalists typically abide by Kant's dictum that one cannot be obliged to do something unless one already has the capacity to do it, with the latter determined (so argue naturalists) by the empirical conditions in which agents find themselves, including their bodies and their social and physical environments. This dictum is meant to militate against supernatural criteria of knowledge and action that would judge humans in terms more appropriate to deities—for example, requiring that one consider all possible consequences or all possible evidence before deciding what to do or what to believe.

This is, of course, my starting-point on this occasion, as it has been on others. In positioning my whole enterprise in relation to the notion of bounded rationality, I have made it clear that how the essential tension is to be understood and managed 'must be constrained by what is empirically known about human beings.' More than that. In emphasizing the importance of articulating norms or describing institutions that are capable of attracting and retaining the allegiance of personnel or participants, I have emphasized the link, not merely between Kant's ought and Kant's can, but, rather, between Rawls's ought and Rawls's will. Let me explain.

As I've pointed out on a previous occasion,[19] norms and institutional structures are defended, within a pragmatist framework (e.g. of Rawls and Rorty), by arguing that these norms can secure 'uptake' on the part of those who are called upon to organize their activities in terms of them. Without such uptake, there is no conformity to the norms. And this is a problem because conformity to the norms is what is supposed to produce the benefits in terms of which the norms themselves are justified. Accordingly, without uptake, no conformity; without conformity, no benefits; without benefits, no justification for conformity.

This is the pragmatist circle of justification, if you will, but what it draws our attention to, I believe, is that norms and institutions, to succeed for human beings, with their infirmities and limitations, must be accessible to human beings as such, where this means not only within their capacities (the point that Kant emphasized), but, more importantly, with their concurrence.

Of course, as Rawls himself already saw, institutions and norms can generate their own support. (This is a generalization of the idea of adaptive preferences.[20]) There is, I think, ample reason to think that this will happen in a well-organized community of enquiry. After all, it provides for asset ownership and engenders an economy of esteem in which participants can understand themselves as contributing to a larger project through their unique individual efforts.

9.5 The issue of scalability

I claim to have identified a number of devices—e.g. multiple accountabilities, the novelty premium—which can be recruited to counteract some of the inhibitors—e.g. social comparison, principal/agency issues—of effective cooperation and collaboration in the pursuit of understanding and improved competence. As Steve Fuller reminds us, however, there are, if you will, *issues of scale* that need to be addressed in evaluating this general approach. He says, in particular:[21]

> Rousseau is famous for confronting the scale-sensitive character of democratic regimes. A true believer in the classical paradigm of democracy, the Athenian polis, Rousseau argued that a democracy lost its 'governing principle' once it reached a size that enabled the formation of conflicting interests. His point was that democracy, in the strict sense, only flourished in small, homogeneous societies, whose members respected each other as equals and were thus willing to abide by a group consensus.

What is the applicability of these general observations to our own case?

First of all, we need to confront the issue of 'directionality,' as it were. Have we described mechanisms which have to scale *up* or, rather, *down* in order to be applicable to and effective in the situations in which knowledge and skills are actually honed?

The answer is not altogether straightforward. First of all, as Karin Knorr Cetina has shown,[22] work in scientific communities, at least, takes place at a variety of different scales, and, as Hinds and Kiesler document,[23] may involve collections whose participants are widely dispersed and may never, in fact, attain consciousness of one another as the members of a group. Secondly, some of the findings on which we have drawn are from small-group psychology (with potential issues of 'ecological validity' when extrapolated to larger, or even just more anonymous settings), whereas others are relevant to precisely these more anonymous, e.g. market, situations with unknown relevance on a smaller scale.

Notwithstanding these difficulties, let's review the main 'devices,' as I have called them and check them for 'scalability.'

Perhaps the key point is that even relative anonymity, or, more properly, 'impersonality,' is likely itself to attenuate the effects of social comparison pressures which are active in small-scale, face-to-face settings (where, in particular, they have been investigated and manipulated). Where individuals do not have to depend on one another for effective joint agency, or where the chains which connect them in some jointly effective collective enterprise are themselves obscure to trace, they will be less subject to the sorts of pressures which otherwise might inhibit them sharing distinctive information or engaging in robust debate.

Of course, we can look at this rather general point in a different way, one which engaged more fully with the specific devices we have been considering. For example, we can view the sort of 'impersonality' that I mentioned as itself an effect, in certain situations, of precisely the multiple accountabilities which individuals find themselves subject to in even moderately complicated and 'true-to-life' circumstances. Similarly, one's dealings with others engaged in the same domain of enquiry become less personal when issues of asset ownership are in play or when, because of divisions of labor, one's primary relations are with individuals within, rather than outside, one's own activity team.

Impersonality is a likely effect too of the 'management by content' scenario which I sketched in Section 6.3. If individuals, including individuals with little face-to-face contact, have their relations with one another mediated by the objects with which they are severally engaged, then their dealings with one another as subjects are likely to be more impersonal. They will, to put it crudely, be oriented to the object, rather than to each other, and thus may find it easier to exchange information and ideas with one another.

Of course, as we have shown, transparency (Section 6.5) is an important element in orienting individuals to one another in an ideally collaborative way. And, certainly, transparency becomes harder to effect in cases where large-scale processes are engaged. Nevertheless, the everyday experience of scholars readily shows how widespread are the rituals of acknowledgement that promote transparency and a sense of shared purpose. I have, for example, already, as I wrote these words, cited about 180 works (excluding 'self-citations'). Of the authors concerned, only eight are known to me personally; the others I know *only* through their writings. Notwithstanding this fact, my acknowledgments are sincere and my sense of dependence on the work of those I cite is profound.

Indeed, there is a kind of double or ironic benefit here. On the one hand, I take seriously a sense of interdependence precisely because I appreciate how the work of others complements and facilitates my own. On the other hand, I can adopt a reasonably impersonal attitude toward the work of others to the degree that they are (relatively speaking) strangers to me and, certainly, not part of any group or even collective where I might actually experience inhibitory social comparison pressures.

And all this, of course, before we invoke novelty premiums, multiple accountabilities, and other mechanisms which are surely at work, on whatever scale I and others might function. I conclude, therefore, that there is no Rouseauvian 'scale problem' when it comes to the collective production of knowledge.

9.6 Philosophy as cartography

If this is a work of epistemology, then it is rather an unusual one, and not, primarily, because it is a work of specifically *social* epistemology. What makes this work different from some, perhaps most, others, even now (and certainly by the standards current when the author was first a postgraduate student), is that this work aims not to dictate the journey, as do many epistemological treatises, but, rather, to map the terrain. There is therefore a strong element of the descriptive, and only, perhaps, a whiff, but, after all, a definite whiff, of the normative. After all, mapping the terrain is not an innocent exercise. If we say that there are hills here or a stream to a certain depth, then, even if we are not telling travelers not to pursue paths in these directions, we are implying something, relevant to their own planning, about their likelihood of success.

This, anyway, is what I hope to have accomplished here. I have tried to draw a map. It shows where be the dragons (social comparison pressures) and what sorts of implements might be needed to beat more congenial lowland paths (accountability, novelty premiums, division of labor, communication, etc.). I would like, in the rest of this chapter, briefly to situate such a project in relation to a *discovery* approach (specifically to the normative) which is more familiar to philosophers and their admirers.

What the philosopher can do who cannot tell people what they should think about a particular topic is to tell them what kinds of things people have thought about that topic, what kinds of ways of thinking people have employed to think about that topic, and what kinds of issues are raised for those who attempt to think about that topic in a systematic way. If the philosopher cannot specify the destination, if you will, that reason must arrive at, she can, anyway, become *a cartographer of the intellectual (and emotional) territory* across which different people or groups of people will find or rather make pathways of their own.[24] She can identify some landmarks, though not all paths will pass them, and she can chart the contours or gradients of thought, though some individuals or groups may prefer steeper than normal and some shallower than normal pathways across these gradients, and, indeed, to some participants, the gradients might seem quite different than they do to the 'theorist'—the 'terrain,' after all, is a territory of contestability, and this can include contestation about what the terrain itself is actually like. The philosopher, on my account, might be able to say what sorts of intellectual and other 'equipment' will serve for the various journeys—what sorts of tools of thought and of social organization may contribute to successful or entertaining or safe or even hazardous journeys and the like, though, again, and again because of contestability and 'finitude,' her story about the kit-bag may be incomplete for or incompatible with some journeys.

An important proviso. Richard Rorty seems to think that we should attempt more, that we should contribute, as he puts it,[25] to 'cultural politics'; that, to translate his ideas into my idiom, we should urge our fellow enquirers to travel a particular route with us and not merely assay the terrain that we might explore. I suspect that I have not always been able to avoid 'cultural politics' in this sense, and I am not quite sure why I do not want to embrace it as the meta-philosophy underpinning my efforts here. Indeed, it is especially puzzling why someone writing about the specifically *social* or *collective* elements of enquiry should shy away from 'legislating' for or at least 'brow-beating' others. But shy away I do—or at least I aspire to do so. It will be enough, I think, to have drawn a map or to have equipped a kit-bag. If the terrain is to be traversed in teams, and the kit-bag is to be raided by groups of people, they will, inevitably, do so differently, and their doing so differently will, inevitably (?) be worth while. This is what I have tried to show, anyway. My task, then, has been a limited one—to chart the possibilities, not prescribe the chosen way.

Perhaps such a task does not seem either noble or ennobling. But whether noble or not, this is a task that is fit for us and for which we are fit. This is something we can do. We know how to chart the intellectual, emotional, institutional, and architectural territories which we have inhabited and through our activities transformed. We have all the skills of history, sociology, psychology, and other disciplines to draw on and all their findings to consider and use.

Of course, we cannot be sure that our charts, should we make them, will define once and for all what the territory looks like. Even this is not within our powers, and for several reasons.[26]

First of all, like the literal cartographer, our powers are limited in relation to the problem we are using them to solve. (See Section 3.1.) Even the literal cartographer, and even when he is dealing with a (relatively) unchanging landscape, cannot portray that landscape in *all* its detail (and he certainly never tries). So, as with all human re-presentational labors, his map is selective and hence incomplete... even when it makes sense, as I will soon say it doesn't in our case, to think of completeness as at least possible in principle or as at least well-defined enough so that we can speak of greater and greater approximations to completeness.

Secondly, however, whether or not this is the case with the literal cartographer, it is certainly the case in our situation: our charts cannot even aim for finality because the very territory which we use them to map is itself changing, and, indeed, is changing in part as a function of our mapping it. (See Section 3.1.4 for more on this idea.) This may also be true in the case of the literal cartographer as well. Once he shows a line of gradients that might not otherwise have been visible to users of his map, he suggests a path that they might beat over that line and, if they do beat it, they thereby render his representation of the territory, which included no such path, obsolete and hence both admitting of and requiring revision. But whether it is or isn't true of the literal cartographer, it is certainly true of the cartographer of ideas or institutions; she no more describes what has already been made by her fellows than that description is itself assimilated into their practices and becomes a guide for re-constructing what has already been made.

Finally, the territory that we are mapping—is this true for the literal cartographer as well?—is a territory of contestability, to use W.B. Gallie's famous epithet[27]... whose significance, I believe, remains, despite the prominence given it by some theorists, still insufficiently acknowledged and assimilated into the core of philosophical activity. (See, for more detail, Section 3.1.5.) The literal cartographer may be fallible and finite and the territory which he charts may be changing and indeed may change as an indirect result of his own activities. The conceptual cartographer faces another challenge which, for all I know, the literal cartographer doesn't or only rarely does face. The challenge is that, precisely as she, the conceptual cartographer, tries to plot the gradients and landmarks and features of the landscape, others in her community are disputing the character of this very landscape. She who wishes to chart the territory of 'public reason,' for example, had better be prepared to find, when she looks at her data, that different contributors to the investigation of this concept have very different notions of what it means and how it functions. Her chart, to capture this disputatious reality, may well end up resembling one of those Escher drawings that makes sense in its various 'parts' (though 'parts' is not quite the right word), but which does not fit together as a coherent whole.[28] In her case, however, this is not her construction; rather, it is her representation of the incoherence of the various constructions of the participants in this particular human enquiry.

What I have done, then, is present a cartography of some collective elements of enquiry. I eschewed telling us how we ought to enquire in order to enquire rightly. All we can say about 'the right' in relation to collective enquiry is that these are the landmarks, gradients, and other features of this conceptual territory. I cannot plausibly aspire to tell people what path to follow through that territory. And, of course, what I have espied about this territory is necessarily: incomplete, because of my and others' finitude and fallibility; outmoded more or less immediately upon formulation because of the continuing constructive endeavors of actual practitioners (whether or not they are influenced by my or other people's meta-level 'discoveries'); and, in any event, unavoidably 'partial' (now in a quasi-political sense) because of the disputatious character of the constructive activities that form and re-form this landscape.

Philosophy is cartography. That is my first point. My second is that, whatever else philosophy might discover about particular conceptual territories, it has—and I have already summarized the main points—discovered three *topological* properties of the maps it makes of these territories... properties which are common, necessarily (?), to all these maps. To repeat: (1) every map is incomplete—all maps share this property; (2) every map is dated the moment it is drafted and perhaps even beforehand; and, finally, (3) every map of specifically human endeavors and practices is a map of a territory of disputation. Philosophy, after all, does make specifically positive or 'factual' findings, and these are they. Perhaps they do not give or even give a hint of 'the meaning of life' or even 'what is truth; what goodness; what justice'... and perhaps they do. Indeed, perhaps these are, precisely, the only credible answers to these age-old questions. They may not seem to have the right form to be answers to such questions. But they might *be* the answers all the same. Or, at least, they might point, back at us, collectively and historically, as constituting, in our very persons and traditions and disputes, the answers to precisely these questions. *We* are truth and goodness and justice. Or, anyway, we make them and we live with the consequences of making them and hence re-make them too from time to time... and there is nothing more (or less) to these (and other) concepts than what we make of them. I repeat: As Dosi puts it, 'the explanation of why something exists rests on how it became what it is.'[29] This is just the historicism of my project, as announced at Section 1.2, of course.

9.7 A cartography of disciplined enquiry

I have mentioned, repeatedly, that assembly bonuses will be realized, if at all, only when a certain balance of forces exists or has been engineered within a particular community of enquiry. Social-dynamic disinhibitors of collective discourse can, I have argued, be counteracted by various institutional arrangements, supporting and being supported by, certain attitudinal, evaluative, and affective habits of effective enquirers.

In effect, I have identified some fundamental topological features of any social organization capable of producing assembly bonuses (in the face of social comparison inhibitions). These are features whose 'subject' is not, primarily, the individual. They are not, typically, of the form, 'The individual enquirer can behave in this way or that and thereby overcome certain inhibiting factors.' They are, rather, at least typically, of the form, 'Structuring social institutions of enquiry in this or that way may, *ceteris paribus*, enable us to attain assembly bonuses in this area of enquiry.' These design elements, then, are elements of specifically social or institutional design.[30] To design communities of enquiry so that individuals (or at least key individuals) are subject to multiple accountabilities (Section 5.1) is, for example, such a principle of institutional design. Building such a feature into our design does not, to put the matter crudely, legislate for individuals. Rather, it tells us what kinds of community structures will ensure or at least raise the probability that individuals will behave in certain kinds of ways which support enquiry (and will do so despite countervailing pressures). The approach taken here, then, is an approach to specifically social epistemology, in the broadest sense, that, unlike the approaches taken, for example, by Alvin Goldman or Philip Kitcher,[31] is collectivist, rather than individualist (as theirs are). In effect, it naturalizes some of the explicitly collectivist work in social epistemology that has been undertaken, for example, by Helen Longino and Miriam Solomon.[32]

Of course, it is one thing to say 'These are the mechanisms which, were they to exist, would counterbalance the forces, should they actually be in play, which might prevent the assembly bonus of collectivized enquiry from being realized in adequate measure' and it is another thing altogether to claim any 'descriptive adequacy' for such an account. I cannot claim, on the basis of the materials and arguments reviewed here, to have established what any given community of enquiry is actually like. Indeed, even a well-functioning community of enquiry might function well by striking some different balance, perhaps involving wholly different mechanisms, between facilitation and inhibition of assembly bonuses. While I have associated my enquiries with the idea of 'naturalizing' epistemology, I cannot claim to have been describing, or indeed even attempting to describe, any actually existing community of enquiry. I therefore cannot say, in particular, how prevalent are the devices which I have identified and how common are the inhibitors which certainly are common in other contexts. That is a matter for another occasion, and, probably, for other investigators, better equipped than I for such a sociological enquiry. One obvious possibility is that such enquiry would uncover a considerable diversity, among different kinds of communities of enquiry (e.g. pursuing different sorts of knowledge or working in different sorts of circumstances), in the ways in which the various inhibitors and facilitators come into play and are balanced. Such a discovery would, of course, be entirely consistent with one of the 'axioms,' if you will, of my whole approach—namely, that we can and should aspire to cartography, rather than to 'moralizing,' if you will... i.e. setting out the 'one best way' in which to organize enquiry. Just as the individual enquirer has to balance competing standards and demands on her attention; just as the various communities of enquiry have to strike a balance between innovation and conservation; just as there are various devices that can be poised in tension with one another to facilitate collectivization of enquiry... so too can we expect there to be a variety of concrete realizations, in 'real-world' conditions, of the devices and approaches which I have here catalogued. Such inconclusiveness, as it will seem to some, is just what we expected. More determinacy is to be found *in* specific, concretely embodied, and historically enacted communities of enquiry; it is not to be found in utterances, however learned (or not), *about* these communities.

Notes

1 Getting it together

1. Martin Kusch, *Knowledge by Agreement: The Programme of Communitarian Epistemology* (Oxford: Clarendon Press, 2002).
2. On which see, for example, John Turk Saunders and Donald F. Henze, *The Private-Language Problem: A Philosophical Dialogue* (New York: Random House, 1967).
3. Kusch, *Knowledge by Agreement*, 175.
4. As in Richard Rorty, *Contingency, Irony, and Solidarity* (Cambridge and New York: Cambridge University Press, 1989).
5. Steve Fuller, *Social Epistemology* (Bloomington, IN & London: Indiana University Press, 1988).
6. Etienne Wenger, *Communities of Practice: Learning, Meaning and Identity* (Cambridge: Cambridge University Press, 1998).
7. Richard Rorty, ed., *The Linguistic Turn* (Chicago: University of Chicago Press, 1967).
8. David Hiley, James Bohman, and Richard Schusterman, eds, *The Interpretive Turn: Philosophy, Science, Culture* (Ithaca: Cornell University Press, 1992).
9. Theodore Schatzki, Karin Knorr Cetina, and Eike von Savigny, eds, *The Practice Turn in Contemporary Theory* (London and New York: Routledge, 2001).
10. Fred D'Agostino, *Incommensurability and Commensuration: The Common Denominator* (Aldershot: Ashgate, 2003).
11. Pierre Bourdieu and Terry Eagleton, 'Doxa and Common Life: An Interview', in *Mapping Ideology*, ed. Slovoj Zizek (London and New York: Verso, 1994), 238.
12. Scott E. Page, *The Difference* (Princeton: Princeton University Press, 2007), Howard Reingold, *Smart Mobs* (Cambridge, MA: Perseus Press, 2002), Cass R. Sunstein, *Infotopia* (New York: Oxford University Press, 2006), James Surowiecki, *The Wisdom of Crowds* (London and New York: Little Brown/Doubleday, 2004).
13. Francis Bacon, *The New Organon, and Related Writings*, ed. Fulton Henry Anderson (Indianapolis: Bobbs-Merrill, 1960).
14. There are some resonances here with Page's idea of the 'toolboxes' which individual practitioners draw on. My notion of technology is, however, more institutional than Page's, emphasizing the specific and concrete ways in which the 'institutional means for coordinating the dynamics of research' have been embodied in social organizations and cultural norms. See Page, *The Difference.*
15. Fred D'Agostino, *Free Public Reason* (New York: Oxford University Press, 1996).
16. Fred D'Agostino, 'Pluralism and Liberalism', in *Handbook of Political Theory*, ed. Gerald F. Gaus and Chandran Kukathas (Thousand Oaks: Sage, 2004).

17. Fred D'Agostino, '"Demographic" Factors in Revolutionary Science: The Wave Model', *Methodology and Science* 26 (1993): 41–52.
18. Cass Sunstein, *Why Societies Need Dissent* (Cambridge, MA: Harvard University Press, 2003), 54.
19. Bruce Ackerman, *We the People, 1: Foundations* (Cambridge, MA: Belknap Press of Harvard University Press, 1991).
20. Thomas Kuhn, *The Structure of Scientific Revolutions*, 2nd edn (Chicago: University of Chicago Press, 1970), 158.
21. Philip Tetlock, 'Coping with Trade-Offs: Psychological Constraints and Political Implications', in *Elements of Reason*, ed. Arthur Lupia, Mathew McCubbins, and Samuel Popkin (Cambridge: Cambridge University Press, 2000).
22. Deborah Perron Tollefsen, 'Group Deliberation, Social Cohesion, and Scientific Teamwork: Is There Room for Dissent?', *Episteme* 4 (2006): 38.
23. Sunstein, *Infotopia* and *Why Societies Need Dissent.*
24. Page, *The Difference*, 13.
25. E.g., most importantly, Philip Kitcher, 'The Division of Cognitive Labor', *Journal of Philosophy* 87 (1990): 5–22.

2 Essential tensions

1. John W. N. Watkins, 'Against "Normal Science"', in *Criticism and the Growth of Knowledge*, ed. Imre Lakatos and Alan Musgrave (Cambridge: Cambridge University Press, 1970), 26.
2. Imre Lakatos, 'Falsification and the Methodology of Scientific Research Programmes', in *Criticism and the Growth of Knowledge*, ed. Imre Lakatos and Alan Musgrave (Cambridge: Cambridge University Press, 1970), 178.
3. Thomas Kuhn, 'The Essential Tension: Tradition and Innovation in Scientific Research' [1959], in *The Essential Tension: Selected Studies in Scientific Tradition and Change*, ed. Thomas Kuhn (Chicago and London: University of Chicago Press, 1977), 227.
4. Ibid.
5. Ibid., 227–8, no. 2.
6. D'Agostino, '"Demographic" Factors in Revolutionary Science: The Wave Model'. See also Section 6.9.
7. See D'Agostino, *Incommensurability and Commensuration*, Section 23.
8. Kuhn, *The Structure of Scientific Revolutions*, 186.
9. Norbert Kerr, Robert MacCoun, and Geoffrey Kramer, 'Bias in Judgment: Comparing Individuals and Groups', *Psychological Review* 103, no. 4 (1996): 696, emphasis added.
10. B. E. Collins and H. Guetzkow, *A Social Psychology of Group Processes for Decision-Making* (New York: Wiley, 1964).
11. Verlin Hinsz, Scott Tinsdale, and David Vollrath, 'The Emerging Conceptualization of Groups as Information Processors', *Psychological Bulletin* 121, no. 1 (1997): 53, emphasis added.
12. For a thorough overview, see, e.g., Jerry M. Suls and Ladd Wheeler, *Handbook of Social Comparison: Theory and Research* (New York: Kluwer Academic/Plenum Publishers, 2000).

13. This is a version, if you will, of a tension, or a dialectic perhaps, identified by Ernest Gellner, who remarked (*Plough, Sword and Book* (Chicago: University of Chicago Press, 1989), 61) in a slightly different but related idiom, that 'logical and social coherence are inversely related'.
14. Garold Stasser and Zachary Birchmeier, 'Group Creativity and Collective Choice', in *Group Creativity: Innovation through Collaboration*, ed. Paul B. Paulus and Bernard Arjan Nijstad (Oxford: Oxford University Press, 2003), 88.
15. As reported, for instance, in William Shadish and Steve Fuller, eds, *The Social Psychology of Science* (New York and London: The Guilford Press, 1994).
16. See for example Herbert A. Simon, *Economics, Bounded Rationality and the Cognitive Revolution* (Aldershot: Edward Elgar, 1992) and James March and Johan P. Olsen, *Ambiguity and Choice in Organizations* (Bergen: Universitetsforlaget, 1976).
17. See Wenger, *Communities of Practice.*
18. My focus on these matters, in subsequent chapters, is one way in which my analysis complements that offered by Scott Page, who largely ignores the conditions in which 'the consistent pattern' might be circumvented. See Page, *The Difference.*
19. Fred D'Agostino, 'The Legacies of John Rawls', *Journal of Moral Philosophy* 1 (2004): 351–67.
20. Stuart Hampshire, *Morality and Conflict* (Oxford: Basil Blackwell, 1983).
21. Karin Knorr Cetina, *Epistemic Cultures: How the Sciences Make Knowledge* (Cambridge: Harvard University Press, 1999), 3.
22. Ibid., 12.
23. William Shadish, Steve Fuller, and Michael Gorman, 'Social Psychology of Science: A Conceptual and Empirical Research Program', in *The Social Psychology of Science*, ed. William Shadish and Steve Fuller (New York and London: The Guilford Press, 1994), 62.
24. Robert Rosenwein, 'Social Influence in Science: Agreement and Dissent in Achieving Scientific Consensus', in *The Social Psychology of Science*, ed. William Shadish and Steve Fuller (New York and London: The Guilford Press, 1994).
25. Fred D'Agostino, 'Rituals of Cosmopolitanism' (Sites of Cosmopolitanism: Citizenship, Aesthetics, Culture, Brisbane, 6–8 July 2005).
26. Terry Connolly and Ken Koput, 'Naturalistic Decision Making and the New Organizational Context', in *Organizational Decision Making*, ed. Zur Shapira (Cambridge: Cambridge University Press, 2002), 288.
27. Pierre Bourdieu, 'The Specificity of the Scientific Field and the Social Conditions of the Progress of Reason' [1975], in *The Science Studies Reader*, ed. Mario Biagiolo (New York and London: Routledge, 1998), 39.
28. Ibid.

3 The facilitative possibilities

1. For interesting and influential general discussion, see Christopher Cherniak, *Minimal Rationality* (Cambridge, MA: The MIT Press, 1986).
2. Hampshire, *Morality and Conflict*, 106.

3. Christopher McMahon, *Collective Rationality and Collective Reasoning* (Cambridge: Cambridge University Press, 2001), 159.
4. David Dequech, 'Bounded Rationality, Institutions, and Uncertainty', *Journal of Economic Issues* 35, no. 4 (2001): 912.
5. For the classic papers, see M. Davis, ed., *The Undecidable: Basic Papers on Undecidable Propositions, Unsolvable Problems and Computable Functions* (Hewlett, NY: Raven Press, 1965).
6. Stuart Kauffman, *At Home in the Universe: The Search for Laws of Self-Organization and Complexity* (New York: Oxford University Press, 1995), 289.
7. See for example Roger Buck, 'Do Reflexive Predictions Pose a Special Problem for the Social Scientist', *Philosophy of Science* 30 (1963): 359–69.
8. See for example Steve Woolgar, ed., *Knowledge and Reflexivity* (London: Sage, 1988).
9. W. B. Gallie, *Philosophy and the Historical Understanding* (London: Chatto and Windus, 1964).
10. For fuller discussion, see D'Agostino, 'Pluralism and Liberalism'.
11. For example Isaiah Berlin, *Four Essays on Liberty* (London: Oxford University Press, 1969).
12. John Gray, *Post-Liberalism: Studies in Political Thought* (London: Routledge, 1993), 65, emphasis added.
13. Berlin, *Four Essays on Liberty*, 1.
14. Luigi Marengo, Giovanni Dosi, Paolo Legreni, and Corrado Pasquali, 'The Structure of Problem-Solving Knowledge and the Structure of Organizations', *Industrial and Corporate Change* 9, no. 4 (2000): 760–1.
15. Moral particularists certainly, and in my view rightly, think that this is quite common. See especially Jonathan Dancy, *Moral Reasons* (Oxford: Blackwell, 1993).
16. Claus Wilkie, *http://en.wikipedia.org/Wiki/Image:Fitness-Landscape-Cartoon.Png* (2001). Accessed 15 March 2009.
17. For the fitness landscape idea, see Kauffman, *At Home in the Universe.*
18. 'The Structure of Problem-Solving Knowledge and the Structure of Organizations', 760.
19. John Rawls, *Political Liberalism* (New York: Columbia University Press, 1993), 55–7.
20. Brian Loasby, *Choice, Complexity and Ignorance* (Cambridge: Cambridge University Press, 1976), 56.
21. See for example Frederick F. Schmitt, ed., *Socializing Epistemology* (Lanham: Rowman and Littlefield, 1994).
22. For representative post-Fuller approaches to social epistemology, see for example Alvin Goldman, *Knowledge in a Social World* (Oxford: Clarendon Press, 1999), Kusch, *Knowledge by Agreement.*
23. Kerr, MacCoun, and Kramer, 'Bias in Judgment: Comparing Individuals and Groups'.
24. See Fred D'Agostino, 'The Aimless Rationality of Science', *International Studies in the Philosophy of Science* 4 (1990): 33–50.
25. Hans Radder, 'Experimental Reproducibility and the Experimenters' Regress', *Proceedings of the Biennial Meetings of the Philosophy of Science Association* 1 (1992): 63–73.

26. Kerr, MacCoun, and Kramer, 'Bias in Judgment: Comparing Individuals and Groups', 696.
27. Sunstein, *Infotopia*, 25.
28. Christian List and Robert E. Goodin, 'Epistemic Democracy: Generalizing the Condorcet Jury Theorem', *Journal of Political Philosophy* 9, no. 3 (2001): 277–306.
29. Of course, as Sunstein points out, the Jury Theorem has what he calls a 'dark side'. If members of the community are less than 50 per cent likely to be correct, then we don't get error reduction, but, rather, error amplification. See Sunstein, *Infotopia*, 28ff.
30. This is *not* the same 'cognitive division of labor' that Kitcher famously surveyed. See Philip Kitcher, 'The Division of Cognitive Labor'. It is, however, closer to the original in Adam Smith, *An Inquiry into the Nature and Causes of the Wealth of Nations* (London: Routledge, 1893).
31. Baruch Fischoff and Stephen Johnson, 'The Possibility of Distributed Decision Making', in *Organizational Decision Making*, ed. Zur Shapira (Cambridge: Cambridge University Press, 2002), 225.
32. See D'Agostino, 'Rituals of Cosmopolitanism'.
33. This is a possibility that is much emphasized by Scott Page, *The Difference*.
34. See for example Rachel Laudan and Larry Laudan, 'Dominance and the Disunity of Method', *Philosophy of Science* 56 (1989): 221–33 and Chun Wei Choo, *The Knowing Organization: How Organizations Use Information to Construct Meaning, Create Knowledge, and Make Decisions* (New York and Oxford: Oxford University Press, 1998).
35. Kerr, MacCoun, and Kramer, 'Bias in Judgment: Comparing Individuals and Groups', 696, emphasis added.
36. S. Moscovici, 'Toward a Theory of Conversion Behavior', in *Advances in Experimental Social Psychology, Vol. 13*, ed. L. Berkowitz (New York: Academic Press, 1980).
37. Carsten De Dreu and Michael West, 'Minority Dissent and Team Innovation: The Importance of Participation in Decision Making', *Journal of Applied Psychology* 86, no. 6 (2001): 1192.
38. McMahon, *Collective Rationality and Collective Reasoning*, 134.
39. Cass Sunstein, 'Deliberating Groups Versus Prediction Markets (or Hayek's Challenge to Habermas)', *Episteme* 3 (2006): 194.
40. See also Charlan Nemeth, 'Minority Dissent as a Stimulant to Group Performance', in *Group Process and Productivity*, ed. Stephen Worchel, Wendy Wood, and Jeffry Simpson (Newbury Park, CA: Sage Publications, 1992) and Thomas Scheidel, 'Divergent and Convergent Thinking in Group Decision-Making', in *Communication and Group Decision Making*, ed. Randy Y. Hirokawa (Thousand Oaks, CA: Sage Publishers, 1996).
41. James Bohman, 'Deliberative Democracy and the Epistemic Benefits of Diversity', *Episteme* 3 (2006): 177.
42. Robert E. Goodin, 'The Epistemic Benefits of Multiple Biased Observers', *Episteme* 3 (2006): 167. Notice how Goodin's account in effect acknowledges some of the mechanisms we have considered above.
43. Ibid., 167–8.
44. McMahon, *Collective Rationality and Collective Reasoning*, Chapter 5, Section 1.

45. Ibid., 110.
46. Garold Stasser, 'Pooling of Unshared Information During Group Discussion', in *Group Process and Productivity*, ed. Stephen Worchel, Wendy Wood, and Jeffry Simpson (Newbury Park, CA: Sage Publications, 1992), 48.
47. Fred D'Agostino, 'Adjudication as an Epistemological Concept', *Synthese* 79 (1989): 230–56.
48. Fred D'Agostino, 'Ethical Pluralism and the Role of Opposition in Democratic Politics', *The Monist* 73 (1990): 437–63 and 'Pluralism and Liberalism'.
49. Philip Kitcher, *Science, Truth, and Democracy* (Oxford: Oxford University Press, 2001), 114.
50. Bourdieu, 'The Specificity of the Scientific Field and the Social Conditions of the Progress of Reason', 39.

4 The consistent pattern

1. Stasser, 'Pooling of Unshared Information During Group Discussion', 49.
2. Stasser and Birchmeier, 'Group Creativity and Collective Choice', 88.
3. Stasser, 'Pooling of Unshared Information During Group Discussion', 49.
4. Sunstein has put a lot of emphasis on this in recent publications, but the present study complements his enquiries in identifying a large number of mechanisms which might be involved in dysfunctional sampling of dispersed knowledge. See for example 'Deliberating Groups Versus Prediction Markets (or Hayek's Challenge to Habermas)', *Infotopia*, and *Why Societies Need Dissent.*
5. Donelson Forsyth, 'Social Comparison and Influence in Groups', in *Handbook of Social Comparison: Theory and Research*, ed. Jerry M. Suls and Ladd Wheeler (New York: Kluwer Academic/Plenum Publishers, 2000), 94.
6. Sunstein, *Why Societies Need Dissent*, 83.
7. De Dreu and West 'Minority Dissent and Team Innovation: The Importance of Participation in Decision Making', 1199.
8. Sara Kiesler and Jonathon N. Cummings, 'What Do We Know About Proximity and Distance in Work Groups? A Legacy of Research', in *Distributed Work*, ed. Pamela Hinds and Sara Kiesler (2002), 59, 62.
9. On the significance of heuristics, see for example Gerd Gigerenzer, *Adaptive Thinking* (Oxford: Oxford University Press, 2000); Thomas Gilovich, Dale Griffin, and Daniel Kahneman, *Heuristics and Biases: The Psychology of Intuitive Judgement* (Cambridge: Cambridge University Press, 2002) and Lakatos, 'Falsification and the Methodology of Scientific Research Programmes'.
10. Jerry M. Suls, 'Opinion Comparison', in *Handbook of Social Comparison*, ed. Jerry M. Suls and Ladd Wheeler (New York: Kluwer Academic/Plenum Publishers, 2000), 116.
11. See for example Karl Popper, *Conjectures and Refutations: The Growth of Scientific Knowledge* (London: Routledge and Kegan Paul, 1963).
12. Kuhn, *The Structure of Scientific Revolutions,* 113.
13. Lakatos put this in terms of the 'negative heuristic' or the 'protective belt', which steered enquiry away from the 'hard core' of the research program. See 'Falsification and the Methodology of Scientific Research Programmes'.

14. Peter Carnevale and Dean Pruitt, 'Negotiation and Mediation', *Annual Review of Psychology* 43, no. 1 (1992): 537.
15. David Teece, 'Design Issues for Innovative Firms: Bureaucracy, Incentives and Industrial Structure', in *The Dynamic Firm: The Role of Technology, Strategy, Organization, and Regions*, ed. Alfred Chandler, Peter Hagstrom, and Orjan Solvell (Oxford: Oxford University Press, 1998), 135f.
16. For example Michael Bratman, 'Shared Cooperative Activity', in *Faces of Intention*, ed. Michael Bratman (Cambridge: Cambridge University Press, 1999).
17. Deborah Perron Tollefsen, 'Group Deliberation, Social Cohesion, and Scientific Teamwork: Is There Room for Dissent?', 43.
18. Ibid.
19. Forsyth, 'Social Comparison and Influence in Groups', 96, emphasis added.
20. Sunstein, *Why Societies Need Dissent*, 86.
21. For important discussion, see W. Brian Arthur, *Increasing Returns and Path Dependence in the Economy* (Ann Arbor: University of Michigan Press, 1994) and Dennis Mueller, *Public Choice II* (Cambridge: Cambridge University Press, 1989).
22. Randall Collins, *The Sociology of Philosophies: A Global Theory of Intellectual Change* (Cambridge, MA: Belknap Press of Harvard University Press, 1998), 38.
23. Xueguang Zhou, 'Organization Decision Making as Rule Following', in *Organizational Decision Making*, ed. Zur Shapira (Cambridge: Cambridge University Press, 2002), 262.
24. Susan Leigh Star, 'The Politics of Formal Representations: Wizards, Gurus, and Organizational Complexity', in *Ecologies of Knowledge*, ed. Susan Leigh Star (Albany: State University of New York Press, 1995), 104.
25. Steve Woolgar, 'Representation, Cognition, and Self: What Hope for an Integration of Psychology and Sociology?', in *Ecologies of Knowledge*, ed. Susan Leigh Star (Albany: State University of New York Press, 1995), 163, emphasis added.
26. McMahon, *Collective Rationality and Collective Reasoning*, 184–5, emphasis added.
27. Xueguang, 'Organization Decision Making as Rule Following', 275.
28. Paul DiMaggio and Walter Powell, 'The Iron Cage Revisited: Institutional Isomorphism and Collective Rationality in Organizational Fields', *American Sociological Review* 48 (1983): 151.
29. Robert Robinson and Laura Kray, 'Naïve Realism and the Search for Social Change and Perceived Legitimacy', in *The Psychology of Legitimacy*, ed. John Jost and Brenda Major (Cambridge: Cambridge University Press, 2001), 151.
30. Steve Fuller, *The Governance of Science* (Philadelphia: The Open University Press, 2000).
31. Bourdieu, 'The Specificity of the Scientific Field and the Social Conditions of the Progress of Reason'.
32. Eric von Hippel, '"Sticky Information" And the Locus of Problem Solving: Implications for Innovation', in *The Dynamic Firm: The Role of Technology, Strategy, Organization, and Regions*, ed. Alfred Chandler, Peter Hagstrom, and Orjan Solvell (Oxford: Oxford University Press, 1998), 61.
33. Ibid., 63, emphasis added.

34. The *locus classicus* is Michael Polanyi, *The Tacit Dimension* (Garden City, NY: Doubleday, 1967); see also Teece, 'Design Issues for Innovative Firms'.
35. See for example Jean-Jacques Laffont, *The Principal Agent Model* (Cheltenham: Edward Elgar, 2003).
36. Matthew J. Grawitch and David C. Munz, 'Individual and Group Affect in Problem-Solving Workgroups', in *Emotions in Organizational Behavior*, ed. Charmine E. Haertel, Wilfred J. Zerbe, and Neal M. Ashkanasy (Mahwah and London: Lawrence Erlbaum Associates, 2005), 127.
37. Ibid., 136–7.
38. James March, 'Understanding How Decisions Happen in Organizations', in *Organizational Decision Making*, ed. Zur Shapira (Cambridge: Cambridge University Press, 2002), 12.
39. D'Agostino, ' "Demographic" Factors in Revolutionary Science: The Wave Model'.

5 Motivating diversity

1. For example Page, *The Difference*; Reingold, *Smart Mobs*; Sunstein, *Infotopia*; and Surowiecki, *The Wisdom of Crowds*.
2. Hinsz Verlin, Scott Tinsdale, and David Vollrath, 'The Emerging Conceptualization of Groups as Information Processors', 45.
3. Helen Longino, *The Fate of Knowledge* (Princeton: Princeton University Press, 2002), 155.
4. See for example Bernard Manin, Adam Przeworski, and Susan Carol Stokes, *Democracy, Accountability, and Representation* (Cambridge and New York: Cambridge University Press, 1999).
5. Shadish, Fuller, and Gorman, 'Social Psychology of Science', 62.
6. The literary theorist Mikhail Bakhtin gave a charming example of this. See Mikhail Mikhailovich Bakhtin, *The Dialogic Imagination: Four Essays* (Austin: University of Texas Press, 1981).
7. Michael Hogg and Deborah Terry, ed., 'Social Identity Theory and Organizational Processes', *Social Identity Processes in Organizational Contexts* (Philadelphia: Psychology Press, 2001), 8.
8. For example Philip Pettit, *Republicanism: A Theory of Freedom and Government* (Oxford: Clarendon Press, 1997).
9. Fuller, *The Governance of Science*, 7.
10. Ibid., 15.
11. See Geoffrey Brennan and Philip Pettit, *The Economy of Esteem* (Oxford: Oxford University Press, 2003).
12. See for example David Shatz, *Peer Review: A Critical Inquiry* (Lanham, MD: Rowman & Littlefield, 2004).
13. McMahon, *Collective Rationality and Collective Reasoning*.
14. Geoffrey and Philip, *The Economy of Esteem*.
15. Sunstein, *Why Societies Need Dissent*, 71.
16. The phrase 'novelty premium' is not in common currency. See however Marcos Chamon, Alejo Costa, and Luca Antonio Ricci, 'Is There a Novelty Premium on New Financial Instruments?', in *International Monetary Fund, IMF Working Papers: 08/109* (2008).

17. John Ziman, *Public Knowledge: The Social Dimension of Science* (Cambridge: Cambridge University Press, 1968), 95.
18. F. A. Hayek, 'The Use of Knowledge in Society', *The American Economic Review* 35, no. 4 (1945): 519–30.
19. Nicolai Foss, ' "Coase Vs Hayek": Economic Organization and the Knowledge Economy', *International Journal of the Economics of Business* 9, no. 1 (2002): 9–17.
20. See for example Ove Granstrand, *The Economics and Management of Intellectual Property* (Cheltenham and Northampton, MA: Edward Elgar, 1999).
21. Rosenwein, 'Social Influence in Science'.
22. P. R. Laughlin, 'Social Combination Processes of Cooperative Problem-Solving Groups on Verbal Intellective Tasks', in *Progress in Social Psychology, Volume 1*, ed. M. Fishbein (Hillsdale, NJ: Erlbaum, 1980).
23. Rosenwein, 'Social Influence in Science', 274.
24. As in D'Agostino, 'Pluralism and Liberalism'.
25. Rosenwein, 'Social Influence in Science', 274.
26. Kuhn, *The Structure of Scientific Revolutions*, 10.
27. Fred D'Agostino, 'Incommensurability and Commensuration', *Studies in the History and Philosophy of Science* 31, no. 3 (2000): 429–47.
28. See for example Fred D'Agostino, 'Kuhn's Risk-Spreading Argument', *Episteme* 1, no. 3 (2005): 201–9 and D'Agostino, 'Rituals of Cosmopolitanism'.
29. Kuhn, *The Structure of Scientific Revolutions*, 173.
30. Forsyth, 'Social Comparison and Influence in Groups', 83.
31. Margit Osterloh and Bruno Frey, 'Motivation, Knowledge Transfer, and Organizational Forms', *Organization Science* 11, no. 5 (2000): 539.
32. Ibid., 540.
33. Ibid.
34. Kuhn, *The Structure of Scientific Revolutions*, 80.
35. Shadish, Fuller, and Gorman, 'Social Psychology of Science', 23–4, emphasis added.
36. Catherine Durnell Cramton, 'The Mutual Knowledge Problem and Its Consequences for Dispersed Collaboration', *Organization Science* 12, no. 3 (2001): 193.
37. This mechanism *extends* the 'classical' account of the fundamental attribution error by eliminating the asymmetry, in the classical account, between the actor and the observer. Typically, actors are situationists about their own behavior, whereas observers of it are more prone to 'the error'. On our account, the actor is also an observer and, accordingly, is liable to this error in her own case. I note that Popper's account (e.g. *The Logic of Scientific Discovery* (London: Hutchinson, 1972)) is closer to the 'classical' account in allowing for the possibility that each individual adopts the actor role with respect to his own activities and the observer role with respect to the activities of others and, accordingly, that the balance between conservation and innovation is sustained socially, through the interaction between actors and observers. This is surely correct. My account adds, however, a layer of self-criticism that is, probably, differentially present across the community.
38. Bohman, 'Deliberative Democracy and the Epistemic Benefits of Diversity', 180–1.

39. Catherine Durnell Cramton, 'Attribution in Distributed Work Groups', in *Distributed Work*, ed. Pamela Hinds and Sara Kiesler (2002), 200.
40. For example in Karl Popper, *The Logic of Scientific Discovery*.
41. Brenda Major and Toni Schmader, 'Legitimacy and the Construal of Social Disadvantage', in *The Psychology of Legitimacy*, ed. John Jost and Brenda Major (Cambridge: Cambridge University Press, 2001), 181.
42. Kuhn, *The Structure of Scientific Revolutions*, 82–3.
43. Deborah Perron Tollefsen, 'Group Deliberation, Social Cohesion, and Scientific Teamwork: Is There Room for Dissent?', 37.

6 A culture of enquiry

1. Scheidel, 'Divergent and Convergent Thinking in Group Decision-Making', 117–18.
2. Susan Leigh Star and James Griesemer, 'Institutional Ecology, "Translations," And Boundary Objects: Amateurs and Professionals in Berkeley's Museum of Vertebrate Zoology, 1907–39', in *The Science Studies Reader*, ed. Mario Biagioli (New York and London: Routledge, 1998), 505.
3. For a concurrent judgment, see Robert Nola, 'Review of Harvey Siegel, *Relativism Refuted*', *British Journal for the Philosophy of Science* 40 (1989): 419–427.
4. Kuhn, *The Structure of Scientific Revolutions*, 185.
5. Ibid., 186.
6. See however D'Agostino, '"Demographic" Factors in Revolutionary Science: The Wave Model', Paul Hoyningen-Huene, *Reconstructing Scientific Revolutions* (Chicago: University of Chicago Press, 1993), Kitcher, 'The Division of Cognitive Labor', and Alexander Rueger, 'Risk and Diversification in Theory Choice', *Synthese* 109 (1996): 263–80.
7. Some theorists reject this implication, i.e. that superiority according to rules of assessment *means* that all rational participants must prefer the superior version *in the sense that* they must work on it. They develop a version of the risk-spreading argument that assumes diversity in individuals' assessments of what variant it would be most beneficial for them individually to work on, given that the assessment of superiority is fallible and hence consistent with another variant being better, long-term, at least potentially. That there could be risk-spreading through this mechanism puts pressure on the assumption that assessment superiority dictates uniform behavior. But it doesn't undermine the main points which I want to make—e.g. that diversity within a community of enquiry is important to risk-spreading and hence effective and efficient exploration of the domain of enquiry. All that happens if we accept the 'cost-benefit' model of individuals' behavior is that the point at which diversity enters the system is changed a bit. This may not be insignificant sociologically. Certainly, we are representing a different *kind* of community of enquiry when we are grounding diversity in (a) its members' different interpretation of shared values, or (b) its members' different assessments of individual advantage.
8. D'Agostino, 'Incommensurability and Commensuration'.

9. Kuhn, *The Essential Tension*, 324.
10. Since this point is true of rules as well as values, the sharp contrast between a values-based and a rules-based approach is actually overdrawn. We might view it as some Kuhnian scaffolding, that we can now dispense with, that enabled him to see the point about diversity of interpretations. (This is the qualification whose necessity I flagged above.)
11. See Ronald Dworkin, *Law's Empire* (London: Fontana Press, 1986), 86 for an argument to this effect in the case of legal interpretation.
12. Barry Barnes, 'Practice as Collective Action', in *The Practice Turn in Contemporary Theory*, ed. Theodore Schatzki, Karin Knorr Cetina, and Eike von Savigny (London: Routledge, 2001), 20.
13. D'Agostino, 'Rituals of Cosmopolitanism'.
14. Fred D'Agostino, 'The Sinews of a Free Society', in *A Passion for Politics: Essays in Honour of Graham Maddox*, ed. Tim Battin (Sydney: Pearson Education Australia, 2005).
15. Laudan and Laudan, 'Dominance and the Disunity of Method'.
16. '"Demographic" Factors in Revolutionary Science: The Wave Model' and *Free Public Reason*. See also Section 6.1.3 below.
17. Sunstein, *Why Societies Need Dissent*, 10.
18. My wave is not quite Sunstein's cascade since the process he describes depends on social comparison pressures, whereas the wave I have in mind depends on the gradual objective improvement of one of the options for choice and the gradual subjective acceptance of that superiority. It depends, if you like, on the development of 'persuasive arguments'; see Section 4.5.
19. Christian List and Philip Pettit, 'Aggregating Sets of Judgments: An Impossibility Result', *Economics and Philosophy* 18, no. 1 (2002): 89–110.
20. Ibid., 95–6.
21. Alvin Goldman, 'Group Knowledge Versus Group Rationality: Two Approaches to Social Epistemology', *Episteme* 1, no. 1 (2004): 13.
22. And it does so without, as List and Pettit might fear, undermining the solidarity of the community of enquiry. That solidarity is founded on a shallow agreement about values, rules, and concrete exemplars, it is not *propositionally* grounded, and hence isn't threatened by any supposed incoherence among the specifically propositional commitments of the community's members. See also Section 6.1.5.
23. Choo Chun Wei, *The Knowing Organization*, xvii.
24. D'Agostino, 'Kuhn's Risk-Spreading Argument', 205.
25. D'Agostino, 'Rituals of Cosmopolitanism'. See also Fuller, *Social Epistemology*, Chapter 9.
26. Loasby, *Choice, Complexity and Ignorance*, 140.
27. Wenger, *Communities of Practice: Learning, Meaning and Identity*, 83–4, emphasis added.
28. Kuhn, *The Structure of Scientific Revolutions*, 186.
29. Ibid.
30. D'Agostino, 'Pluralism and Liberalism'.
31. Kitcher, 'The Division of Cognitive Labor'.
32. Susan and Griesemer 'Institutional Ecology, "Translations," And Boundary Objects', 506.
33. Kuhn, *The Structure of Scientific Revolutions*, 10.

34. Hampshire, *Morality and Conflict*, 106.
35. This is, in effect, an instance of the 'frame problem', as it's called in artificial intelligence, i.e. the problem of specifying, in advance, what is and what isn't relevant collateral information that bears on a particular problem. See for example Zenon Pylyshyn, ed., *The Robot's Dilemma: The Frame Problem in Artificial Intelligence* (Norwood, NJ: Ablex Publishing Corporation, 1987).
36. Kuhn, *The Structure of Scientific Revolutions*, 44.
37. I've shown A and B as sharing an element of their interpretations in order to make explicit what should be obvious: their interpretations can't differ 'too much' without difficulties for the dynamics of the community to which they belong. Even if A and B are both talking about X, they really aren't talking about the same thing if everything A says about X differs from anything that B says about X. In this case, they might begin to think that there wasn't really any particularly good reason to continue to address one another. A's 'image' of X is *so* different from B's, that they might as well keep their ideas to themselves.
38. I have chosen the word 'solidarity' here to mark my general indebtedness to the ideas of Richard Rorty, another, more esteemed, explicator of Kuhn. See for example Rorty, *Contingency, Irony, and Solidarity*.
39. In a different context, Friedrich Hayek argued that freedom is better protected in a regime of proscriptive rules. See F. A. Hayek, *Law, Legislation and Liberty* (Chicago: University of Chicago Press, 1973–6).
40. Lakatos. 'Falsification and the Methodology of Scientific Research Programmes', in *Criticism and the Growth of Knowledge*, 135, emphasis added.
41. Knorr Cetina, *Epistemic Cultures*, 171.
42. Donald Norman, *The Design of Everyday Things* (London: The MIT Press, 1998), 9ff, 131ff.
43. Knorr Cetina, *Epistemic Cultures*, 173–4.
44. W. M. Cohen and D. A. Levinthal, 'Absorptive Capacity', *Administrative Science Quarterly* 35 (1990): 128–52.
45. De Dreu and West, 'Minority Dissent and Team Innovation', 1192.
46. Fred D'Agostino, 'Rituals of Impartiality', *Social Theory and Practice* 27 (2001): 65–81.
47. Elizbeth A. Mannix, Terri Griffith, and Margaret A. Neale, 'The Phenomenology of Conflict in Distributed Work Teams', in *Distributed Work*, ed. Pamela Hinds and Sara Kiesler (2002), 213–14.
48. Ibid.
49. Ibid., 214.
50. Ibid., 215.
51. Kuhn, *The Structure of Scientific Revolutions*, 177.
52. The locus classicus is M. Friedman and L. P. Savage, 'The Utility Analysis of Choices Involving Risk', *Journal of Political Economy* 56 (1948): 279–304.
53. John Payne, 'The Scarecrow's Search: A Cognitive Psychologist's Perspective on Organization Decision Making', in *Organizational Decision Making*, ed. Zur Shapira (Cambridge: Cambridge University Press, 2002), 364.
54. I note that my analysis diverges from the analysis which Kuhn himself provides, especially in recognizing that there need not, as Kuhn seems to have

thought, be a tight connection, empirically, between seniority and conservatism. See Kuhn, *The Structure of Scientific Revolutions*, 147ff. I note too that, although my analysis trades on the observations of March and Shapira, it shows some significant parallels with Philip Kitcher's analysis of 'Individual Responses to Innovation'. See Philip Kitcher, *The Advancement of Science* (New York: Oxford University Press, 1993).

55. See, on Donald Campbell's analysis of the role of generational change in relation to epistemic norms, Shadish, Fuller, and Gorman, 'Social Psychology of Science'.
56. Pierre Bourdieu, 'The Forms of Capital', in *Handbook for Theory and Research for the Sociology of Education*, ed. J.G. Richardson (New York: Greenwood Press, 1986).
57. See also, for a slightly different version of this story, Kitcher, *Science, Truth, and Democracy*, 112–13. See also, reporting some ideas of Frederick Grinnell, *The Scientific Attitude*, 2nd ed. (New York: Guilford Press, 1992), Miriam Solomon, 'Norms of Epistemic Diversity', *Episteme* 4 (2006): 25.
58. Bourdieu, 'The Specificity of the Scientific Field and the Social Conditions of the Progress of Reason', 36.
59. Richard Moreland and John Levine, 'Problem Identification in Groups', in *Group Process and Productivity*, ed. Stephen Worchel, Wendy Wood, and Jeffry Simpson (Newbury Park, CA: Sage Publications, 1992), 25.
60. Ibid.
61. D'Agostino, *Free Public Reason*, Section 34.
62. Charles Taylor, *Sources of the Self: The Making of the Modern Identity* (Cambridge: Cambridge University Press, 1989), 4.
63. For a more sophisticated analysis, using some tools of economics, and with allusions to 'bandwagon' and 'herding' models, see Jesus Zamora Bonilla, 'The Elementary Economics of Scientific Consensus', *Theoria* 14, no. 3 (1999): 461–88.

7 The division of labor in communities of enquiry

1. See for example Kitcher, *The Advancement of Science*, 'The Division of Cognitive Labor', and *Science, Truth, and Democracy*.
2. Kitcher, *The Advancement of Science*, 303.
3. Kuhn, *The Structure of Scientific Revolutions*, 186.
4. Smith, *An Inquiry into the Nature and Causes of the Wealth of Nations*, Book I, Chapter 1.
5. Karl Marx, *Capital* (London: Lawrence and Wishart, 1954), vol. I, Chapter 14.
6. Rueger, 'Risk and Diversification in Theory Choice'.
7. Kitcher, 'The Division of Cognitive Labor'.
8. See for example Carliss Baldwin and Kim Clark, *Design Rules, Volume 1. The Power of Modularity* (Cambridge, MA: The MIT Press, 2000); Stefano Brusoni and Andrea Prencipe, 'The Division of Labor and the Division of Knowledge', in *Dynamics of Industry and Innovation: Organizations, Networks and Systems* (Copenhagen, Denmark: DRUID Tenth Annual Summer Conference, 2005); Giovanni Dosi, 'Technological Paradigms and Technological Trajectories', *Research Policy* 11 (1982): 147–62; Sendil Ethiraj and Daniel

Levinthal, 'Modularity and Innovation in Complex Systems', *Management Science* 50, no. 2 (2004): 159–73; Koen Frenken, *Innovation, Evolution and Complexity Theory* (Cheltenham & Northampton, MA: Edward Elgar, 2006); Stuart A. Kauffman, *The Origins of Order: Self Organization and Selection in Evolution* (New York: Oxford University Press, 1993); Luigi Marengo and Giovanni Dosi, 'Division of Labor, Organizational Coordination and Market Mechanisms in Collective Problem-Solving', *Journal of Economic Behavior & Organization* 58 (2005): 303–26; Marengo *et al.*, 'The Structure of Problem-Solving Knowledge and the Structure of Organizations', in *Managing in the Modular Age: Architectures, Network, and Organizations*, ed. Garud Raghu, Arun Kumaraswamy, and Richard Laglois (Malden, MA and Oxford: Blackwell Publishing, 2003); Ron Sanchez and Joseph Mahoney, 'Modularity, Flexibility, and Knowledge Management in Product and Organization Design', *Strategic Management Journal* 17 (1996): 63–76; and Eric Von Hippel, 'Task Partitioning: An Innovation Process Variable', *Research Policy* 19 (1990): 407–18.

9. See for example Allen Newell and Herbert A. Simon, *Human Problem Solving* (Englewood Cliffs, NJ: Prentice-Hall, 1972) and Herbert A. Simon, *The Sciences of the Artificial* (Cambridge: M.I.T. Press, 1969).
10. See for example Berlin, *Four Essays on Liberty* and Gray, *Post-Liberalism: Studies in Political Thought*. See also D'Agostino, 'Ethical Pluralism and the Role of Opposition in Democratic Politics' and 'Pluralism and Liberalism'. Finally, see Section 3.1.
11. See also Marengo and Dosi, 'Division of Labor, Organizational Coordination and Market Mechanisms in Collective Problem-Solving', 309.
12. D'Agostino, *Incommensurability and Commensuration.*
13. Frenken, *Innovation, Evolution and Complexity Theory*, 8, emphasis added.
14. It is worth noting that such complex relations among evaluative standards are recognized, or at least asserted, by ethical 'particularists' such as Jonathon Dancy; indeed, Dancy's particularism is grounded, precisely, in such complexity, not that, to my knowledge, he puts it that way himself. See Dancy, *Moral Reasons* (Oxford: Blackwell, 1993).
15. Ethiraj and Levinthal, 'Modularity and Innovation in Complex Systems', 161, emphasis added.
16. See Lee Altenberg, 'NK Fitness Landscapes', in *The Handbook of Evolutionary Computation*, ed. T. Back, D. Fogel, and Z. Michalewicz (Oxford: Oxford University Press, 1997) and Kauffman, *The Origins of Order: Self Organization and Selection in Evolution*.
17. John Payne, James Bettman, and David Schkade, 'Measuring Constructed Preferences: Towards a Building Code', *Journal of Risk and Uncertainty* 19, no. 1 (1999): 243–59.
18. Frenken, *Innovation, Evolution and Complexity Theory*, 33–4.
19. P. Saviotti and J. Metcalfe, 'A Theoretical Approach to the Construction of Technological Output Indicators', *Research Policy* 14 (1984): 141–51.
20. Marengo *et al.*, 'The Structure of Problem-Solving Knowledge and the Structure of Organizations', 760, emphasis added.
21. Frenken, *Innovation, Evolution and Complexity Theory*, 15.
22. Ibid.
23. Ibid., 14.

24. It is worth remarking, of course, that the key features of a rugged evaluative landscape—many mediocre local optima—might ground a case for Simonian 'satisficing'. Why bother to expend resources in searching a space where a local peak is not likely to differ very much from any other local peak. Find a peak and stay there might be rather good advice. For an interesting discussion see Ibid., 25ff.
25. Ibid., 20.
26. Marengo and Dosi, 'Division of Labor, Organizational Coordination and Market Mechanisms in Collective Problem-Solving', 309, emphasis added.
27. Frenken, *Innovation, Evolution and Complexity Theory*, 21.
28. Even relative to qualifications I will review later, this approach encounters a potential difficulty which has not, to my knowledge, been discussed in the literature. The difficulty arises because or to the extent that the technically achievable sub-division of the *object* of enquiry maps properly onto the decomposition of the *evaluative* space. The suggestion is that it cannot be guaranteed that the ways in which it is possible to decompose the object of enquiry will 'line up' in the right sort of way with the ways in which it is possible to partition the values interdependencies. (I am not sure how this relates to what Saviotti calls (Saviotti and Metcalfe, 'A Theoretical Approach to the Construction of Technological Output Indicators', 66), according to Frenken, the '*imaging pattern* between technical and service characteristics'.)
29. Baldwin and Clark, *Design Rules, Volume 1*.
30. Ibid., 50.
31. Marengo and Dosi, 'Division of Labor, Organizational Coordination and Market Mechanisms in Collective Problem-Solving', 309.
32. Ibid., 310.
33. Baldwin and Clark, *Design Rules, Volume 1*, 50.
34. Raghu, Kumaraswamy, and Laglois, eds, *Managing in the Modular Age*, 2.
35. Connolly and Koput, 'Naturalistic Decision Making and the New Organizational Context', 288.
36. Kuhn, *The Structure of Scientific Revolutions*, 17.
37. See in particular Baldwin and Clark, *Design Rules, Volume 1*; Ethiraj and Levinthal, 'Modularity and Innovation in Complex Systems'; Sanchez and Mahoney, 'Modularity, Flexibility, and Knowledge Management in Product and Organization Design'.
38. Sanchez and Mahoney, 'Modularity, Flexibility, and Knowledge Management in Product and Organization Design', 65.
39. Compare Ethiraj and Levinthal, 'Modularity and Innovation in Complex Systems', 161 and Baldwin and Clark, *Design Rules, Volume 1*, 14.
40. I do not mean to be dismissive of communication as an effective response to complexity and will return to this topic in Section 7.5.
41. It is not a trivial matter to partition a system into subsystems. I consider this issue later. It is also appropriate to recall my earlier discussion of the potential mismatch between feasible partitions of the technical characteristics of the object of enquiry and partitions of the service characteristics by which that object is to be judged. Especially when the problem space is constructed as we go, enquirers are not working directly with service, but rather, with technical characteristics. Any sensible partitioning of the task is probably therefore going to be made on a technical basis.

42. Baldwin and Clark, *Design Rules, Volume 1.*
43. I have adopted a way of describing these decisions and the design rules to which they lead, despite the way in which they are commonly portrayed, as decisions which enforce partitions on technical characteristics, rather than service characteristics. This only adds to my anxieties about the potential for a mismatch. There is no reason to believe, *a priori*, that technically feasible subdivisions will track evaluatively modular partitions.
44. Loasby, *Choice, Complexity and Ignorance*, 77–8.
45. Baldwin and Clark, *Design Rules, Volume 1*, 68–9.
46. Another way of putting this is to say that, in a modularized system, as opposed to a decomposable system, it cannot be guaranteed that the globally optimum system is what you get when you put together the locally optimized subsystems.
47. Ethiraj and Levinthal, 'Modularity and Innovation in Complex Systems', 161, emphasis added.
48. Baldwin and Clark, *Design Rules, Volume 1*, 70.
49. Ethiraj and Levinthal, 'Modularity and Innovation in Complex Systems', 169–70.
50. Marengo *et al.*, 'The Structure of Problem-Solving Knowledge and the Structure of Organizations', 777.
51. Brian Loasby, 'Decision Premises, Decision Cycles and Decomposition', *Industrial and Corporate Change* 9, no. 4 (2000): 722.
52. Kuhn, *The Structure of Scientific Revolutions*, 186.
53. Marengo *et al.*, 'The Structure of Problem-Solving Knowledge and the Structure of Organizations', 799–800.
54. See Raghu, Kumaraswamy, and Laglois, eds, *Managing in the Modular Age: Architectures, Network, and Organizations*, 4 for relevant discussion.
55. Giovanni Dosi, 'Opportunities, Incentives, and the Collective Patterns in Technological Change', *Economic Journal* 107 (1997): 1531.
56. Marengo *et al.*, 'The Structure of Problem-Solving Knowledge and the Structure of Organizations', 757.
57. Bourdieu, 'The Specificity of the Scientific Field and the Social Conditions of the Progress of Reason', 39.
58. Michael Polanyi, 'The Republic of Science', *Minerva* 1 (1962): 55.
59. See D'Agostino, *Free Public Reason.*
60. See my account of Rawls's pure proceduralism. Ibid.
61. Theodore Porter, *Trust in Numbers* (Princeton: Princeton University Press, 1995) provides an account of some of the difficulties that might be attendant on this.
62. This need not determine the other two to join her at this location. Perhaps they want to hedge their bets, guarding against the possibility that there is, somewhere else in the large terrain, a summit which is even higher. Of course, even if they decide this, each has enough information to know better than to move right from A or left from C. In this interval, they have already identified the optimum and an even better optimum can only be found by moving left from A and right from C.
63. Kauffman, *The Origins of Order*, 252.

8 Some communities of enquiry

1. See also Hayek, *Law, Legislation and Liberty*.
2. Hayek, 'The Use of Knowledge in Society', 526.
3. On this point see William Butos and Thomas McQuade, 'Mind, Market and Institutions: The Knowledge Problem in Hayek's Thought'.
4. See for example Sunstein, *Infotopia* and Michael Abramowicz, 'Information Markets, Administrative Decisionmaking, and Predictive Cost-Benefit Analysis', *University of Chicago Law Review* 71, no. 3 (2004): 933–1020.
5. Tom Bell, 'Prediction Markets for Promoting the Progress of Science and the Useful Arts', *George Mason Law Review* 14, no. 1 (2006): 47.
6. Ibid.
7. It is interesting that Sunstein provides a clear exposition of the hidden profile issue (Sunstein, 'Deliberating Groups Versus Prediction Markets (or Hayek's Challenge to Habermas)', 199) but seems to fail to see how, in an information market, or prediction market as he calls it, the case in which the hidden profile is not likely to be revealed deliberatively is also one in which that profile (and its implications for judgment) are not likely to be revealed through market mechanisms. Indeed, he seems not to see the significance of precisely the distinction I have been laboring—that between using knowledge in a market and disclosing that knowledge. He says for example (Ibid., 205.), that 'because people stand to gain or lose from their investments [in an information market], they have a strong incentive to use (and in that sense to disclose) whatever private information they hold'.
8. Ibid., 201.
9. Ibid., 207–8.
10. D'Agostino, *Incommensurability and Commensuration*, Section 21.
11. Bohman, 'Deliberative Democracy and the Epistemic Benefits of Diversity', 187.
12. Alexander Hamilton, James Madison, and John Jay, *The Federalist* (Washington, DC: National Home Library Foundation, 1938), Sections 62–3.
13. Fuller, *Social Epistemology*, 6.
14. Mohammadreza Hojat, Joseph Gonnella, and Addeane Caelleigh, 'Impartial Judgment by the "Gatekeepers" of Science: Fallibility and Accountability in the Peer Review Process', *Advances in Health Sciences Education* 8 (2003): 75–96.
15. Kuhn, *The Structure of Scientific Revolutions*.
16. D'Agostino, *Incommensurability and Commensuration*.
17. Lorraine Daston, 'Objectivity and the Escape from Perspective' [1992], in *The Science Studies Reader*, ed. Mario Biagiolo (New York and London: Routledge, 1999), 117–18.
18. Porter, *Trust in Numbers* and D'Agostino, *Incommensurability and Commensuration*.
19. Moreland and Levine, 'Problem Identification in Groups'.
20. Connolly and Koput, 'Naturalistic Decision Making and the New Organizational Context', 288.
21. Detlof von Winterfeldt and Ward Edwards, *Decision Analysis and Behavioral Research* (Cambridge: Cambridge University Press, 1986), 26.

22. Marengo *et al.*, 'The Structure of Problem-Solving Knowledge and the Structure of Organizations', 760–1.
23. Kuhn, *The Structure of Scientific Revolutions*, 13ff.
24. Lakatos, 'Falsification and the Methodology of Scientific Research Programmes', 157.
25. Connolly and Koput, 'Naturalistic Decision Making and the New Organizational Context', 287.
26. Robert Goodin ('The Epistemic Benefits of Multiple Biased Observers') has suggested another reason why individuals might disperse themselves over the fitness landscape—namely, that a scientist's chances of enhancing her reputation are better if she commits herself to a *distinctive* position.
27. There is some tension, of course, with my points, made earlier, about the way in which we make such a problem landscape up as we go.

9 Naturalizing the essential tension

1. Zamora Bonilla, 'The Elementary Economics of Scientific Consensus'.
2. Richard Rorty, *Objectivity, Relativism and Truth* (Cambridge: Cambridge University Press, 1991), 83.
3. Joseph Rouse, *Knowledge and Power* (Ithaca, NY: Cornell University Press, 1987).
4. See D'Agostino, *Free Public Reason* and John Rawls, *A Theory of Justice* (Oxford: Oxford University Press, 1973).
5. D'Agostino, 'Adjudication as an Epistemological Concept' and 'The Aimless Rationality of Science'.
6. D'Agostino, 'The Legacies of John Rawls'.
7. See, for discussion, Fred D'Agostino, 'Teleology, Value, and the Foundations of Scientific Method', *Methodology and Science* 24 (1991): 119–34.
8. Broadly evolutionary approaches—the methods must be truth-conducive because they have survived in competition with other methods—presuppose all sorts of things which are false. See for argument Fred D'Agostino, 'Transcendence and Conversation: Two Conceptions of Objectivity', *American Philosophical Quarterly* 30 (1993): 87–108.
9. D'Agostino, *Free Public Reason* and *Incommensurability and Commensuration.*
10. Kenneth Knight, 'Questions About Matrix Management', in *Matrix Management*, ed. Kenneth Knight (Farnborough: Gower Press, 1977), 114–15.
11. Ibid., 114.
12. Ibid.
13. HLA Hart, *The Concept of Law* (Oxford: Clarendon Press, 1962).
14. Knight, 'Questions About Matrix Management'.
15. Ibid.
16. Ibid.
17. Ibid.
18. Fuller, *The Governance of Science*, 37.
19. D'Agostino, 'The Legacies of John Rawls'.
20. On which see, for example, Jon Elster, *Sour Grapes: Studies in the Subversion of Rationality* (Cambridge and New York: Cambridge University Press, 1983).
21. Fuller, *The Governance of Science*, 38.

22. Knorr Cetina, *Epistemic Cultures*.
23. Pamela Hinds and Sara Kiesler, eds, *Distributed Work* (Cambridge: The MIT Press, 2002).
24. For the use of geographical tropes by Nietzsche and by Gilles Deleuze and Felix Guattari, *A Thousand Plateaus: Capitalism and Schizophrenia* (Minneapolis, Minn.: University of Minnesota Press, 1987), see Stephan Günzel, 'Nietzsche's Geophilosophy', *Journal of Nietzsche Studies* 25 (2003): 78–91. There is obvious resonance here with the 'landscape' and 'path' terminology of Chapter 7.
25. Richard Rorty, *Philosophy as Cultural Politics: Philosophical Papers Volume 4* (Cambridge: Cambridge University Press, 2007).
26. Fred D'Agostino, 'Epistemology of the Social Sciences', in *A Companion to Epistemology*, ed. Jonathan Dancy and Ernest Sosa (Oxford: Basil Blackwell, 1992).
27. Gallie, *Philosophy and the Historical Understanding*.
28. M. C. Escher, *The Graphic Work of M.C. Escher* (New York: Hawthorn Books, 1960).
29. Dosi, 'Opportunities, Incentives, and the Collective Patterns in Technological Change', 1531.
30. See Robert E. Goodin, ed., *The Theory of Institutional Design* (Cambridge: Cambridge University Press, 1996).
31. See for example Alvin Goldman, 'Veritistic Social Epistemology' (paper presented at the Twentieth World Congress of Philosophy, Charlottesville, 2000) and Kitcher, *Science, Truth, and Democracy*.
32. Longino, *The Fate of Knowledge* and Solomon, 'Norms of Epistemic Diversity'.

References

Abramowicz, Michael. 'Information Markets, Administrative Decisionmaking, and Predictive Cost-Benefit Analysis.' *University of Chicago Law Review* 71, no. 3 (2004): 933–1020.

Ackerman, Bruce. *We the People, 1: Foundations*. Cambridge, MA: Belknap Press of Harvard University Press, 1991.

Altenberg, Lee. 'NK Fitness Landscapes.' In *The Handbook of Evolutionary Computation*, edited by T. Back, D. Fogel and Z. Michalewicz. Oxford: Oxford University Press, 1997.

Arthur, W. Brian. *Increasing Returns and Path Dependence in the Economy*. Ann Arbor: University of Michigan Press, 1994.

Bacon, Francis. *The New Organon, and Related Writings*. Edited by Fulton Henry Anderson. Indianapolis: Bobbs-Merrill, 1960.

Bakhtin, Mikhail Mikhailovich. *The Dialogic Imagination: Four Essays*. Austin: University of Texas Press, 1981.

Baldwin, Carliss, and Kim Clark. *Design Rules, Volume 1: The Power of Modularity*. Cambridge, MA: The MIT Press, 2000.

Barnes, Barry. 'Practice as Collective Action.' In *The Practice Turn in Contemporary Theory*, edited by Theodore Schatzki, Karin Knorr Cetina and Eike von Savigny. London: Routledge, 2001.

Bell, Tom. 'Prediction Markets for Promoting the Progress of Science and the Useful Arts.' *George Mason Law Review* 14, no. 1 (2006): 37–92.

Berlin, Isaiah. *Four Essays on Liberty*. London: Oxford University Press, 1969.

Blissett, Marlan. *Politics in Science*. Boston: Little, Brown and Company, 1972.

Bohman, James. 'Deliberative Democracy and the Epistemic Benefits of Diversity.' *Episteme* 3 (2006): 175–91.

Bourdieu, Pierre. 'The Forms of Capital.' In *Handbook for Theory and Research for the Sociology of Education*, edited by J. G. Richardson. New York: Greenwood Press, 1986.

———. 'The Specificity of the Scientific Field and the Social Conditions of the Progress of Reason.' In *The Science Studies Reader*, edited by Mario Biagiolo, 31–50. New York and London: Routledge, 1998 [1975].

Bourdieu, Pierre, and Terry Eagleton. 'Doxa and Common Life: An Interview.' In *Mapping Ideology*, edited by Slovoj Zizek, 265–77. London and New York: Verso, 1994.

Bratman, Michael. 'Shared Cooperative Activity.' In *Faces of Intention*, edited by Michael Bratman, 93–109. Cambridge: Cambridge University Press, 1999.

Brennan, Geoffrey, and Philip Pettit. *The Economy of Esteem*. Oxford: Oxford University Press, 2003.

Brusoni, Stefano, and Andrea Prencipe. 'The Division of Labor and the Division of Knowledge.' In *Dynamics of Industry and Innovation: Organizations, Networks and Systems*. Copenhagen, Denmark: DRUID Tenth Annual Summer Conference, 2005.

Buck, Roger. 'Do Reflexive Predictions Pose a Special Problem for the Social Scientist.' *Philosophy of Science* 30 (1963): 359–69.

Butos, W. N., and T. J. McQuade. 'Mind, Market, and Institutions: The Knowledge Problem in Hayek's Thought.' In *F. A. Hayek as a Political Economist*, edited by J. Birner, P. Garrouste and T. Aimar. London and New York: Routledge, 2002.

Carnevale, Peter, and Dean Pruitt. 'Negotiation and Mediation.' *Annual Review of Psychology* 43, no. 1 (1992): 531ff.

Chamon, Marcos, Alejo Costa, and Luca Antonio Ricci. 'Is There a Novelty Premium on New Financial Instruments?', In *International Monetary Fund, IMF Working Papers: 08/109*, 2008.

Cherniak, Christopher. *Minimal Rationality*. Cambridge, MA: The MIT Press, 1986.

Choo, Chun Wei. *The Knowing Organization: How Organizations Use Information to Construct Meaning, Create Knowledge, and Make Decisions*. New York and Oxford: Oxford University Press, 1998.

Cohen, W. M., and D. A. Levinthal. 'Absorptive Capacity: A New Perspective on Learning and Innovation.' *Administrative Science Quarterly* 35 (1990): 128–52.

Collins, B. E., and H. Guetzkow. *A Social Psychology of Group Processes for Decision-Making*. New York: Wiley, 1964.

Collins, Randall. *The Sociology of Philosophies: A Global Theory of Intellectual Change*. Cambridge, MA: Belknap Press of Harvard University Press, 1998.

Connolly, Terry, and Ken Koput. 'Naturalistic Decision Making and the New Organizational Context.' In *Organizational Decision Making*, edited by Zur Shapira. Cambridge: Cambridge University Press, 2002.

Cramton, Catherine Durnell. 'Attribution in Distributed Work Groups.' In *Distributed Work*, edited by Pamela Hinds and Sara Kiesler. Cambridge: MIT Press, 2002.

———. 'The Mutual Knowledge Problem and Its Consequences for Dispersed Collaboration.' *Organization Science* 12, no. 3 (2001): 346ff.

D'Agostino, Fred. 'Adjudication as an Epistemological Concept.' *Synthese* 79 (1989): 231–56.

———. 'The Aimless Rationality of Science.' *International Studies in the Philosophy of Science* 4 (1990): 33–50.

———. '"Demographic" Factors in Revolutionary Science: The Wave Model.' *Methodology and Science* 26 (1993): 41–52.

———. 'Epistemology of the Social Sciences.' In *A Companion to Epistemology*, edited by Jonathan Dancy and Ernest Sosa, 479–83. Oxford: Basil Blackwell, 1992.

———. 'Ethical Pluralism and the Role of Opposition in Democratic Politics.' *The Monist* 73 (1990): 437–63.

———. *Free Public Reason*. New York: Oxford University Press, 1996.

———. 'From the Organization to the Division of Cognitive Labor.' *Politics, Philosophy and Economics* 8 (2009): 101–29.

———. 'Incommensurability and Commensuration: Lessons from (and to) Ethico-Political Theory.' *Studies in the History and Philosophy of Science* 31, no. 3 (2000): 429–47.

———. *Incommensurability and Commensuration: The Common Denominator*. Aldershot: Ashgate, 2003.

———. 'Kuhn's Risk-Spreading Argument.' *Episteme* 1, no. 3 (2005): 201–09.

———. 'The Legacies of John Rawls.' *Journal of Moral Philosophy* 1 (2004): 351–67.

———. 'Naturalizing the Essential Tension.' *Synthese* 162 (2008): 275–308.
———. 'Pluralism and Liberalism.' In *Handbook of Political Theory*, edited by Gerald F. Gaus and Chandran Kukathas. Thousand Oaks: Sage, 2004.
———. 'Rituals of Cosmopolitanism.' Paper presented at the conference on Sites of Cosmopolitanism: Citizenship, Aesthetics, Culture, Brisbane, 6–8 July 2005.
———. 'Rituals of Impartiality.' *Social Theory and Practice* 27 (2001): 65–81.
———. 'The Sinews of a Free Society: Autonomy, Democracy and Education.' In *A Passion for Politics: Essays in Honour of Graham Maddox*, edited by Tim Battin, 99–109. Sydney: Pearson Education Australia, 2005.
———. 'Teleology, Value, and the Foundations of Scientific Method.' *Methodology and Science* 24 (1991): 119–34.
———. 'Transcendence and Conversation: Two Conceptions of Objectivity.' *American Philosophical Quarterly* 30 (1993): 87–108.
Dancy, Jonathan. *Moral Reasons*. Oxford: Blackwell, 1993.
Daston, Lorraine. 'Objectivity and the Escape from Perspective.' In *The Science Studies Reader*, edited by Mario Biagiolo, 110–23. New York and London: Routledge, 1999 [1992].
Davis, M., ed. *The Undecidable: Basic Papers on Undecidable Propositions, Unsolvable Problems and Computable Functions*. Hewlett, NY: Raven Press, 1965.
De Dreu, Carsten, and Michael West. 'Minority Dissent and Team Innovation: The Importance of Participation in Decision Making.' *Journal of Applied Psychology* 86, no. 6 (2001): 1191–201.
Deleuze, Gilles, and Felix Guattari. *A Thousand Plateaus: Capitalism and Schizophrenia*. Minneapolis, MN: University of Minnesota Press, 1987.
Dequech, David. 'Bounded Rationality, Institutions, and Uncertainty.' *Journal of Economic Issues* 35, no. 4 (2001): 911ff.
DiMaggio, Paul, and Walter Powell. 'The Iron Cage Revisited: Institutional Isomorphism and Collective Rationality in Organizational Fields.' *American Sociological Review* 48 (1983): 147–60.
Dosi, Giovanni. 'Opportunities, Incentives, and the Collective Patterns in Technological Change.' *Economic Journal* 107 (1997): 1530–47.
———. 'Technological Paradigms and Technological Trajectories.' *Research Policy* 11 (1982): 147–62.
Dworkin, Ronald. *Law's Empire*. London: Fontana Press, 1986.
Elster, Jon. *Sour Grapes: Studies in the Subversion of Rationality*. Cambridge and New York: Cambridge University Press, 1983.
Escher, M. C. *The Graphic Work of M. C. Escher*. New and rev. ed. New York: Hawthorn Books, 1960.
Ethiraj, Sendil, and Daniel Levinthal. 'Modularity and Innovation in Complex Systems.' *Management Science* 50, no. 2 (2004): 159–73.
Fischoff, Baruch, and Stephen Johnson. 'The Possibility of Distributed Decision Making.' In *Organizational Decision Making*, edited by Zur Shapira. Cambridge: Cambridge University Press, 2002.
Forsyth, Donelson. 'Social Comparison and Influence in Groups.' In *Handbook of Social Comparison: Theory and Research*, edited by Jerry M. Suls and Ladd Wheeler. New York: Kluwer Academic/Plenum Publishers, 2000.
Foss, Nicolai. '"Coase Vs Hayek": Economic Organization and the Knowledge Economy.' *International Journal of the Economics of Business* 9, no. 1 (2002): 9ff.
Frenken, Koen. *Innovation, Evolution and Complexity Theory*. Cheltenham & Northampton MA: Edward Elgar, 2006.

Friedman, M., and L. P. Savage. 'The Utility Analysis of Choices Involving Risk.' *Journal of Political Economy* 56 (1948): 279–304.
Fuller, S. *The Governance of Science*. Philadelphia: The Open University Press, 2000.
———. *Social Epistemology*. Bloomington, IN and London: Indiana University Press, 1988.
Gallie, W. B. *Philosophy and the Historical Understanding*. London: Chatto and Windus, 1964.
Gellner, E. *Plough, Sword and Book*. Chicago: University of Chicago Press, 1989.
Gigerenzer, Gerd. *Adaptive Thinking: Rationality in the Real World*. Oxford: Oxford University Press, 2000.
Gilovich, Thomas, Dale Griffin, and Daniel Kahneman. *Heuristics and Biases: The Psychology of Intuitive Judgement*. Cambridge: Cambridge University Press, 2002.
Goldman, Alvin. 'Group Knowledge Versus Group Rationality: Two Approaches to Social Epistemology.' *Episteme* 1, no. 1 (2004): 11–22.
———. *Knowledge in a Social World*. Oxford: Clarendon Press, 1999.
———. 'Veritistic Social Epistemology.' Paper presented at the Twentieth World Congress of Philosophy, Charlottesville 2000.
Goodin, Robert E. 'The Epistemic Benefits of Multiple Biased Observers.' *Episteme* 3 (2006): 166–74.
———, ed. *The Theory of Institutional Design*. Cambridge: Cambridge University Press, 1996.
Granstrand, Ove. *The Economics and Management of Intellectual Property: Towards Intellectual Capitalism*. Cheltenham and Northampton, MA: E. Elgar, 1999.
Grawitch, Matthew J., and David C. Munz. 'Individual and Group Affect in Problem-Solving Workgroups.' In *Emotions in Organizational Behavior*, edited by Charmine E. Haertel, Wilfred J. Zerbe and Neal M. Ashkanasy, 119–42. Mahwah and London: Lawrence Erlbaum Associates, 2005.
Gray, John. *Post-Liberalism: Studies in Political Thought*. London: Routledge, 1993.
Grinnell, Frederick. *The Scientific Attitude*. 2nd edn. New York: Guilford Press, 1992.
Günzel, Stephan. 'Nietzsche's Geophilosophy.' *Journal of Nietzsche Studies* 25 (2003): 78–91.
Hamilton, Alexander, James Madison, and John Jay. *The Federalist*. Sesquicentennial ed. with an introduction by Edward Mead Earle. Washington, DC: National Home Library Foundation, 1938.
Hampshire, Stuart. *Morality and Conflict*. Oxford: Basil Blackwell, 1983.
Hart, HLA. *The Concept of Law*. Oxford: Clarendon Press, 1962.
Hayek, F. A. *Law, Legislation and Liberty*. Chicago: University of Chicago Press, 1973–6.
———. 'The Use of Knowledge in Society.' *The American Economic Review* 35, no. 4 (1945): 519–30.
Hiley, David, James Bohman, and Richard Schusterman, eds. *The Interpretive Turn: Philosophy, Science, Culture*. Ithaca: Cornell University Press, 1992.
Hinds, Pamela, and Sara Kiesler, eds. *Distributed Work*. Cambridge: The MIT Press, 2002.
Hinsz, Verlin, Scott Tinsdale, and David Vollrath. 'The Emerging Conceptualization of Groups as Information Processors.' *Psychological Bulletin* 121, no. 1 (1997): 43–64.
Hogg, Michael, and Deborah Terry. 'Social Identity Theory and Organizational Processes.' In *Social Identity Processes in Organizational Contexts*, edited

by Michael Hogg and Deborah Terry, 1–12. Philadelphia: Psychology Press, 2001.

Hojat, Mohammadreza, Joseph Gonnella, and Addeane Caelleigh. 'Impartial Judgment by the "Gatekeepers" of Science: Fallibility and Accountability in the Peer Review Process.' *Advances in Health Sciences Education* 8 (2003): 75–96.

Hoyningen-Huene, Paul. *Reconstructing Scientific Revolutions*. Chicago: University of Chicago Press, 1993.

Kauffman, Stuart. *At Home in the Universe: The Search for Laws of Self-Organization and Complexity*. New York: Oxford University Press, 1995.

———. *The Origins of Order: Self Organization and Selection in Evolution*. New York: Oxford University Press, 1993.

Kerr, Norbert, Robert MacCoun, and Geoffrey Kramer. 'Bias in Judgment: Comparing Individuals and Groups.' *Psychological Review* 103, no. 4 (1996): 687–719.

Kiesler, Sara, and Jonathon N. Cummings. 'What Do We Know About Proximity and Distance in Work Groups? A Legacy of Research.' In *Distributed Work*, edited by Pamela Hinds and Sara Kiesler, 2002.

Kitcher, Philip. *The Advancement of Science: Science without Legend, Objectivity without Illusions*. New York: Oxford University Press, 1993.

———. 'The Division of Cognitive Labor.' *Journal of Philosophy* 87 (1990): 5–22.

———. *Science, Truth, and Democracy*. Oxford: Oxford University Press, 2001.

Knight, Kenneth. 'Questions About Matrix Management.' In *Matrix Management*, edited by Kenneth Knight, 109–18. Farnborough: Gower Press, 1977.

Knorr Cetina, Karin. *Epistemic Cultures: How the Sciences Make Knowledge*. Cambridge: Harvard University Press, 1999.

Kuhn, Thomas. *The Essential Tension*. Chicago: University of Chicago Press, 1977.

———. 'The Essential Tension: Tradition and Innovation in Scientific Research.' [1959] In *The Essential Tension: Selected Studies in Scientific Tradition and Change*, edited by Thomas Kuhn. Chicago and London: University of Chicago Press, 1977.

———. *The Structure of Scientific Revolutions*. 2nd edn. Chicago: University of Chicago Press, 1970.

Kusch, Martin. *Knowledge by Agreement: The Programme of Communitarian Epistemology*. Oxford: Clarendon Press, 2002.

Laffont, Jean-Jacques. *The Principal Agent Model: The Economic Theory of Incentives*. Cheltenham: Edward Elgar, 2003.

Lakatos, Imre. 'Falsification and the Methodology of Scientific Research Programmes.' In *Criticism and the Growth of Knowledge*, edited by Imre Lakatos and Alan Musgrave. Cambridge: Cambridge University Press, 1970.

Lakatos, Imre, and Alan Musgrave, eds. *Criticism and the Growth of Knowledge*. Cambridge: Cambridge University Press, 1970.

Laudan, Rachel, and Larry Laudan. 'Dominance and the Disunity of Method.' *Philosophy of Science* 56 (1989): 221–33.

Laughlin, P. R. 'Social Combination Processes of Cooperative Problem-Solving Groups on Verbal Intellective Tasks.' In *Progress in Social Psychology, Volume 1*, edited by M. Fishbein. Hillsdale, NJ: Erlbaum, 1980.

List, Christian, and Robert E. Goodin. 'Epistemic Democracy: Generalizing the Condorcet Jury Theorem.' *Journal of Political Philosophy* 9, no. 3 (2001): 277–306.

List, Christian, and Philip Pettit. 'Aggregating Sets of Judgments: An Impossibility Result.' *Economics and Philosophy* 18, no. 1 (2002): 89–110.

Loasby, Brian. *Choice, Complexity and Ignorance*. Cambridge: Cambridge University Press, 1976.

———. 'Decision Premises, Decision Cycles and Decomposition.' *Industrial and Corporate Change* 9, no. 4 (2000): 709–31.

Longino, Helen. *The Fate of Knowledge*. Princeton: Princeton University Press, 2002.

Major, Brenda, and Toni Schmader. 'Legitimacy and the Construal of Social Disadvantage.' In *The Psychology of Legitimacy*, edited by John Jost and Brenda Major. Cambridge: Cambridge University Press, 2001.

Manin, Bernard, Adam Przeworski, and Susan Carol Stokes. *Democracy, Accountability, and Representation*. Cambridge and New York: Cambridge University Press, 1999.

Mannix, Elizbeth A., Terri Griffith, and Margaret A. Neale. 'The Phenomenology of Conflict in Distributed Work Teams.' In *Distributed Work*, edited by Pamela Hinds and Sara Kiesler. Cambridge: MIT Press, 2002.

March, James. 'Understanding How Decisions Happen in Organizations.' In *Organizational Decision Making*, edited by Zur Shapira. Cambridge: Cambridge University Press, 2002.

March, James Gardner, and Johan P. Olsen. *Ambiguity and Choice in Organizations*. Bergen: Universitetsforlaget, 1976.

Marengo, Luigi, and Giovanni Dosi. 'Division of Labor, Organizational Coordination and Market Mechanisms in Collective Problem-Solving.' *Journal of Economic Behavior & Organization* 58 (2005): 303–26.

Marengo, Luigi, Giovanni Dosi, Paolo Legreni, and Corrado Pasquali. 'The Structure of Problem-Solving Knowledge and the Structure of Organizations.' *Industrial and Corporate Change* 9, no. 4 (2000): 757–88.

Marx, Karl. *Capital*. London: Lawrence and Wishart, 1954.

McMahon, Christopher. *Collective Rationality and Collective Reasoning*. Cambridge: Cambridge University Press, 2001.

Moreland, Richard, and John Levine. 'Problem Identification in Groups.' In *Group Process and Productivity*, edited by Stephen Worchel, Wendy Wood and Jeffry Simpson. Newbury Park, CA: Sage Publications, 1992.

Moscovici, S. 'Toward a Theory of Conversion Behavior.' In *Advances in Experimental Social Psychology, Vol. 13*, edited by L. Berkowitz. New York: Academic Press, 1980.

Mueller, Dennis. *Public Choice II*. Cambridge: Cambridge University Press, 1989.

Nemeth, Charlan. 'Minority Dissent as a Stimulant to Group Performance.' In *Group Process and Productivity*, edited by Stephen Worchel, Wendy Wood and Jeffry Simpson. Newbury Park, CA: Sage Publications, 1992.

Newell, Allen, and Herbert A. Simon. *Human Problem Solving*. Englewood Cliffs, NJ: Prentice-Hall, 1972.

Nisbett, R. and Ross, L. *Human Inference: Strategies and Shortcomings of Social Judgment*. Englewood Cliffs, NJ: Prentice-Hall, 1980.

Nola, Robert. 'Review of Harvey Siegel, *Relativism Refuted*.' *British Journal for the Philosophy of Science* 40 (1989): 419–27.

Norman, Donald. *The Design of Everyday Things*. London: The MIT Press, 1998.

Osterloh, Margit, and Bruno Frey. 'Motivation, Knowledge Transfer, and Organizational Forms.' *Organization Science* 11, no. 5 (2000): 538–50.
Page, Scott E. *The Difference: How the Power of Diversity Creates Better Groups, Firms, Schools, and Societies*. Princeton: Princeton University Press, 2007.
Payne, John. 'The Scarecrow's Search: A Cognitive Psychologist's Perspective on Organization Decision Making.' In *Organizational Decision Making*, edited by Zur Shapira. Cambridge: Cambridge University Press, 2002.
Payne, John, James Bettman, and David Schkade. 'Measuring Constructed Preferences: Towards a Building Code.' *Journal of Risk and Uncertainty* 19, no. 1 (1999): 243ff.
Pettit, P. *Republicanism: A Theory of Freedom and Government*. Oxford: Clarendon Press, 1997.
Polanyi, Michael. 'The Republic of Science.' *Minerva* 1 (1962): 54–73.
———. *The Tacit Dimension*. Garden City, NY: Doubleday, 1967.
Popper, K. *The Logic of Scientific Discovery*. London: Hutchinson, 1972.
———. *Conjectures and Refutations: The Growth of Scientific Knowledge*. London: Routledge and Kegan Paul, 1963.
Porter, Theodore. *Trust in Numbers*. Princeton: Princeton University Press, 1995.
Pylyshyn, Zenon, ed. *The Robot's Dilemma: The Frame Problem in Artificial Intelligence*. Norwood, NJ: Ablex Publishing Corporation, 1987.
Radder, Hans. 'Experimental Reproducibility and the Experimenters' Regress.' *Proceedings of the Biennial Meetings of the Philosophy of Science Association* 1 (1992): 63–73.
Raghu, Garud, Arun Kumaraswamy, and Richard Laglois, eds. *Managing in the Modular Age: Architectures, Network, and Organizations*. Malden, MA and Oxford: Blackwell Publishing, 2003.
Rawls, John. *Political Liberalism*. New York: Columbia University Press, 1993.
———. *A Theory of Justice*. Oxford: Oxford University Press, 1973.
Reingold, Howard. *Smart Mobs*. Cambridge, MA: Perseus Press, 2002.
Robinson, Robert, and Laura Kray. 'Naïve Realism and the Search for Social Change and Perceived Legitimacy.' In *The Psychology of Legitimacy*, edited by John Jost and Brenda Major. Cambridge: Cambridge University Press, 2001.
Rorty, Richard. *Contingency, Irony, and Solidarity*. Cambridge and New York: Cambridge University Press, 1989.
———. *Objectivity, Relativism and Truth*. Cambridge: Cambridge University Press, 1991.
———. *Philosophy and the Mirror of Nature*. Princeton: Princeton University Press, 1979.
———. *Philosophy as Cultural Politics: Philosophical Papers Volume 4*. Cambridge: Cambridge University Press, 2007.
———, ed. *The Linguistic Turn*. Chicago: University of Chicago Press, 1967.
Rosenwein, Robert. 'Social Influence in Science: Agreement and Dissent in Achieving Scientific Consensus.' In *The Social Psychology of Science*, edited by William Shadish and Steve Fuller. New York and London: The Guilford Press, 1994.
Rouse, Joseph. *Knowledge and Power*. Ithaca, NY: Cornell University Press, 1987.
Rueger, Alexander. 'Risk and Diversification in Theory Choice.' *Synthese* 109 (1996): 263–80.

Sanchez, Ron, and Joseph Mahoney. 'Modularity, Flexibility, and Knowledge Management in Product and Organization Design.' *Strategic Management Journal* 17 (1996): 63–76.

Saunders, John Turk, and Donald F. Henze. *The Private-Language Problem: A Philosophical Dialogue*. New York: Random House, 1967.

Saviotti, P., and J. Metcalfe. 'A Theoretical Approach to the Construction of Technological Output Indicators.' *Research Policy* 14 (1984): 141–51.

Schatzki, Theodore, Karin Knorr Cetina, and Eike von Savigny, eds. *The Practice Turn in Contemporary Theory*. London and New York: Routledge, 2001.

Scheidel, Thomas. 'Divergent and Convergent Thinking in Group Decision-Making.' In *Communication and Group Decision Making*, edited by Randy Y. Hirokawa. Thousand Oaks, CA: Sage Publishers, 1996.

Schmitt, Frederick F., ed. *Socializing Epistemology: The Social Dimensions of Knowledge*. Lanham: Rowman and Littlefield, 1994.

Shadish, William, and Steve Fuller, eds. *The Social Psychology of Science*. New York and London: The Guilford Press, 1994.

Shadish, William, Steve Fuller, and Michael Gorman. 'Social Psychology of Science: A Conceptual and Empirical Research Program.' In *The Social Psychology of Science*, edited by William Shadish and Steve Fuller. New York and London: The Guilford Press, 1994.

Shatz, David. *Peer Review: A Critical Inquiry*. Lanham, MD: Rowman & Littlefield, 2004.

Simon, Herbert A. *Economics, Bounded Rationality and the Cognitive Revolution*. Aldershot: Edward Elgar, 1992.

———. *The Sciences of the Artificial*. Cambridge: MIT Press, 1969.

Smith, Adam. *An Inquiry into the Nature and Causes of the Wealth of Nations*. London: Routledge, 1893.

Solomon, Miriam. 'Norms of Epistemic Diversity.' *Episteme* 4 (2006): 23–36.

Star, Susan Leigh. 'The Politics of Formal Representations: Wizards, Gurus, and Organizational Complexity.' In *Ecologies of Knowledge: Work and Politics in Science and Technology*, edited by Susan Leigh Star, 88–118. Albany: State University of New York Press, 1995.

Star, Susan Leigh, and James Griesemer. 'Institutional Ecology, "Translations," And Boundary Objects: Amateurs and Professionals in Berkeley's Museum of Vertebrate Zoology, 1907–39.' In *The Science Studies Reader*, edited by Mario Biagioli, 505–24. New York and London: Routledge, 1998.

Stasser, Garold. 'Pooling of Unshared Information During Group Discussion.' In *Group Process and Productivity*, edited by Stephen Worchel, Wendy Wood and Jeffry Simpson. Newbury Park, CA: Sage Publications, 1992.

Stasser, Garold, and Zachary Birchmeier. 'Group Creativity and Collective Choice.' In *Group Creativity: Innovation through Collaboration*, edited by Paul B. Paulus and Bernard Arjan Nijstad. Oxford: Oxford University Press, 2003.

Suls, Jerry M. 'Opinion Comparison.' In *Handbook of Social Comparison*, edited by Jerry M. Suls and Ladd Wheeler. New York: Kluwer Academic/Plenum Publishers, 2000.

Suls, Jerry M., and Ladd Wheeler. *Handbook of Social Comparison: Theory and Research*. New York: Kluwer Academic/Plenum Publishers, 2000.

Sunstein, Cass. 'Deliberating Groups Versus Prediction Markets (or Hayek's Challenge to Habermas).' *Episteme* 3 (2006): 192–213.

———. *Infotopia: How Many Minds Produce Knowledge*. New York: Oxford University Press, 2006.

———. *Why Societies Need Dissent*. Cambridge, MA: Harvard University Press, 2003.

Surowiecki, James. *The Wisdom of Crowds: Why the Many Are Smarter Than the Few and How Collective Wisdom Shapes Business, Economies, Societies, and Nations*. London and New York: Little Brown/Doubleday 2004.

Taylor, Charles. *Sources of the Self: The Making of the Modern Identity*. Cambridge: Cambridge University Press, 1989.

Teece, David. 'Design Issues for Innovative Firms: Bureaucracy, Incentives and Industrial Structure.' In *The Dynamic Firm: The Role of Technology, Strategy, Organization, and Regions*, edited by Alfred Chandler, Peter Hagstrom and Orjan Solvell, 134–65. Oxford: Oxford University Press, 1998.

Tetlock, Philip. 'Coping with Trade-Offs: Psychological Constraints and Political Implications.' In *Elements of Reason*, edited by Arthur Lupia, Mathew McCubbins and Samuel Popkin, 239–63. Cambridge: Cambridge University Press, 2000.

Tollefsen, Deborah Perron. 'Group Deliberation, Social Cohesion, and Scientific Teamwork: Is There Room for Dissent?', *Episteme* 4 (2006): 37–51.

von Hippel, Eric. '"Sticky Information" And the Locus of Problem Solving: Implications for Innovation.' In *The Dynamic Firm: The Role of Technology, Strategy, Organization, and Regions*, edited by Alfred Chandler, Peter Hagstrom and Orjan Solvell, 60–77. Oxford: Oxford University Press, 1998.

———. 'Task Partitioning: An Innovation Process Variable.' *Research Policy* 19 (1990): 407–18.

von Winterfeldt, Detlof, and Ward Edwards. *Decision Analysis and Behavioral Research*. Cambridge: Cambridge University Press, 1986.

Watkins, John W. N. 'Against "Normal Science".' In *Criticism and the Growth of Knowledge*, edited by Imre Lakatos and Alan Musgrave. Cambridge: Cambridge University Press, 1970.

Wenger, E. *Communities of Practice: Learning, Meaning and Identity*. Cambridge: Cambridge University Press, 1998.

Wilkie, Claus. http://en.wikipedia.org/Wiki/Image:Fitness-Landscape-Cartoon.Png 2001, Accessed 15 March 2009.

Woolgar, Steve. 'Representation, Cognition, and Self: What Hope for an Integration of Psychology and Sociology?', In *Ecologies of Knowledge: Work and Politics in Science and Technology*, edited by Susan Leigh Star, 154–79. Albany: State University of New York Press, 1995.

———, ed. *Knowledge and Reflexivity: New Frontiers in the Sociology of Knowledge*. London: Sage, 1988.

Zamora Bonilla, Jesus. 'The Elementary Economics of Scientific Consensus.' *Theoria* 14, no. 3 (1999): 461–88.

Zhou, Xueguang. 'Organization Decision Making as Rule Following.' In *Organizational Decision Making*, edited by Zur Shapira. Cambridge: Cambridge University Press, 2002.

Ziman, John. *Public Knowledge: The Social Dimension of Science*. Cambridge: Cambridge University Press, 1968.

Index

Note: locators with an *f* or *ff* denote following folios.

accountability, 65, 159, 171
Ackerman, 5
adaptive preferences, 176
affect, 61*f*
Akerlof, 57
Altenberg, 115
assembly bonus, 7*f*, 14, 17*f*, 33, 37, 39*f*, 42, 44*ff*, 59, 64, 71, 100–4, 143, 148, 152, 182*f*
asset ownership, 71*f*, 143, 173*f*
attitudinal, 18, 73*f*, 76, 87

Bacon, 3
Baldwin, 122, 125, 129, 131
Bell, 145*f*
benchmarking, 9, 66
Berlin, 27, 114
Birchmeier, 15, 43
Blisset, 1, 15
Bohman, 38, 77, 148
Bonilla, 165
bounded rationality, 20*f*, 47, 49, 61, 120, 125, 134, 154
Bourdieu, 2, 19, 41, 58, 106*f*, 136
Bratman, 50*f*
Brennan, 70
burdens of judgment, 30

Carnevale, 49
cartography, 10, 17*f*, 32, 60, 178–84
cascade, 5, 8, 87, 108*f*, 116, 128, 146, 195
Choo, 91
Clark, 122, 125, 129, 131
Cohen, 101
collectivism, 2, 4, 6, 20*f*, 30*ff*, 40, 151, 162, 183
Collins, 14, 53, 76*f*
community of enquiry, 2, 5, 9, 11, 16*f*, 59, 62, 64, 74, 81*f*, 86*f*, 91*f*, 98–101, 106, 108, 110, 113, 139, 141*f*, 145, 148*ff*, 154, 159, 161, 165*ff*, 182
complexity, 7, 9, 27, 29, 31*f*, 113*f*, 116*ff*, 126, 131, 133*f*, 137*ff*, 141, 157*f*, 162, 164
Condorcet's Jury Theorem, 33, 189
Connolly, 126, 155, 157
constructed preferences, 117
contestability, 4, 7, 26*f*, 31*f*, 40, 179*f*
Cramton, 77
cultural capital, 68, 106, 151, 154*f*, 174
cultural politics, 179
Cummings, 46

Daston, 153*f*
De Dreu, 37, 46, 100*f*
Dequech, 22*f*
design rule, 129, 131, 133*f*, 140
DiMaggio, 56
discursive paradox, 89*f*
division of labor, 9, 34, 59, 100*ff*, 125, 128, 134, 136*f*, 141*ff*, 162–5, 172, 177
dominance, 35, 87, 89*f*, 96, 110, 159, 165
Dorner, 157
Dosi, 50, 121, 123*f*, 135, 182

economy of esteem, 66*f*, 69, 75, 173
Edwards, 155
epistasis, 115*f*, 121
equilibrium traps, 55, 57, 162*f*
essential tension, 6, 9, 11*f*, 44, 51, 68, 80, 92, 102, 154, 162*f*, 167
exhaustive search, 119*ff*, 126, 131*f*, 140
expectancy disconfirmation, 14, 36, 39, 48*ff*, 73*f*, 82, 153, 163

federalist model, 133, 147–50
Feyerabend, 134
first mover, 8, 53*ff*, 57, 69, 162, 165
Fischoff, 35
fitness landscape, 29, 31, 58, 115, 118, 159

Forsyth, 52*f*, 74
Foss, 72
Foucault, 3
frame problem, 196
Frenken, 115, 117–20
Frey, 75
Fuller, 2, 32, 57, 67, 70, 151, 175*f*
fundamental attribution error, 8, 76*ff*, 163, 193

Gallie, 26, 180
gate keepers, 151
generative entrenchment, 10, 53*ff*, 68, 165
Goldman, 89, 183
Goodin, 38
Grawitch, 61*f*
Gray, 27, 114
Griesemer, 82
Guetzkow, 14

Hampshire, 17, 20, 94
Hart, 171
Hayek, 70, 144, 151
heuristics, 47, 49, 64
hidden profile, 43*f*, 145*f*, 148, 201
Hiley, 2
Hinds, 177
Hinsz, 15, 42, 65
Hippel, 58*f*, 125
historicism, 3, 5, 10, 182
Hogg, 66

incommensurability, 3
inexhaustibility, 20, 23, 31*f*, 61, 94*f*
information market, 143–7
information stickiness, 8, 58*ff*, 64, 145
intellective, 18, 73, 87
interface, 127*ff*, 134
intrinsic motivation, 75*f*

Johnson, 35
joint agency, 50*f*, 60, 65, 87, 101, 172, 177

Kant, 175
Kauffman, 23*ff*, 115, 118, 140*f*
Kerr, 14*f*, 33, 36
Kiesler, 46, 177
Kitcher, 41, 93, 111*f*, 121, 133, 136, 183
Knight, 169–75
Knorr Cetina, 18, 98*ff*, 176
Koput, 126, 155, 157
Kray, 56
Kuhn, 2, 5*f*, 9–13, 15*f*, 49, 51, 63, 66, 72, 74, 76, 80*f*, 83, 85*ff*, 90, 92*ff*, 96, 99, 104–8, 111, 113*f*, 126, 133*f*, 150, 152, 157, 160, 162*f*, 165*f*, 171*f*
Kusch, 1*f*

Lakatos, 11, 92, 97, 152, 157, 160, 171
Laudan, 87
Laughlin, 73
Levine, 107*f*, 155, 157
Levinthal, 101, 115*f*, 130*f*
liberal approach, 138, 152, 165
List, 89
Loasby, 31*f*, 91, 129, 133, 137
Longino, 65, 183
loss aversion, 56

Madison, 150
Mahoney, 128
Major, 78*f*
management by content, 98*f*
Mannix, 102*ff*, 110
March, 16, 56, 63, 106
Marengo, 28*f*, 31, 118, 121, 123*f*, 131, 133*ff*, 138, 156
Marx, 112
McMahon, 22, 37, 40, 54, 69
Mead, 77
Merton, 57, 70
Metcalfe, 117
modularity, 127–32, 200
Moreland, 107*f*, 155, 157
Moscovici, 37
multiple accountabilities, 8, 18, 65, 110, 113, 149, 162*f*, 165, 167, 171, 178, 182
Munz, 61*f*
myopic search, 119*ff*, 126, 129, 132, 158

Nash, 66
negative heuristic, 97, 140
normal science, 79
Norman, 98*f*
novelty premium, 8, 18, 67, 69, 71, 110, 113, 143, 159, 162*f*, 165, 167, 174, 178

Olsen, 56
Orwell, 54
Osterloh, 75

Page, 2, 9, 185
paradigm, 5, 35, 49*f*, 66, 70, 74, 84, 86, 93–6, 98*f*, 106*ff*, 140, 148*ff*, 153, 155, 157*f*, 163
parallel search, 123, 127
particularism, 198
patches, 34, 154
path dependence, 7, 9*f*, 30, 49, 53, 57, 113, 123*f*, 130, 133*ff*, 140, 162*f*, 166*f*
Payne, 106
persuasive arguments, 52, 58, 143
Pettit, 67, 70, 89
pluralism, 3, 8, 10, 41, 73, 86, 93, 113, 136
Polanyi, 136
pool of reasons, 40, 54
Popper, 11, 48, 71, 78, 141, 156
population approach, 133, 148
Porter, 154
positive heuristic, 97
positive test strategy, 71
Powell, 56
principal/agency, 60*f*, 65, 71
private language argument, 1, 33
Pruitt, 49
public reason, 4

Raghu, 124
Rawls, 30, 168, 175
reflexivity, 25*f*
Reingold, 2
relationship conflict, 102*ff*
replication, 32*f*
republicanism, 67*ff*, 138*ff*, 158, 163, 165
residual divergence, 90*f*
risk-aversion, 105, 107, 155, 159, 173*f*
risk-spreading, 6, 9, 16, 18, 83–7, 90*f*, 93*f*, 96, 98, 107, 110*f*, 136, 162*ff*, 194
Robinson, 56
robustness, 38
Rorty, 1*f*, 164, 166, 175, 179
Rosenwein, 73*f*
Rouse, 166
Rousseau, 176
Rueger, 112

sampling dynamics, 44*f*
sampling strategy, 47
Sanchez, 128
satisficing, 199
Saviotti, 117
scalability, 176*ff*
Scheidel, 82
Schmader, 78*f*
secularism, 175*f*
Shadish, 18, 66, 76
shallow consensus, 18, 91
Shapira, 106
Simon, 16, 113, 121
Smith, 112
social comparison, 7, 15–18, 45, 54*f*, 57, 60, 64*f*, 67, 73, 100*f*, 110, 113, 141, 143, 145, 148*f*, 162, 167, 170, 182
social epistemology, 2, 6, 9, 11, 15
Solomon, 183
Star, 53, 82
Stasser, 15, 41, 43
Suls, 47
Sunstein, 2, 4, 8, 33, 37, 45, 52, 70, 87, 109, 146*f*, 150, 201
Surowieckie, 2

tacit knowledge, 60
task conflict, 102*ff*
Taylor, 109
Teece, 50
Terry, 66
Tetlock, 5
Tollefsen, 8, 50*ff*
transparency, 104, 177*f*

undecidability, 23*ff*
uptake, 17, 47, 175

values, 81*ff*

Watkins, 11
wave model, 4, 11, 88, 108, 195
Wenger, 2, 92
West, 37, 45, 100*f*
Wimsatt, 53
Winterfeldt, 155
Woolgar, 53, 55

Zhou, 53, 55*f*
Ziman, 70